Environmental Impact Assessment

Environmental Impact Assessment

Judith Rosales

Delve
PUBLISHING
www.delvepublishing.com

Environmental Impact Assessment

Judith Rosales

Delve Publishing
224 Shoreacres Road
Burlington, ON L7L 2H2
Canada
www.delvepublishing.com

Email: orders@arclereducation.com

ISBN: 978-1-77407-692-7 (Hardcover)

ABOUT THE AUTHOR

Judith Rosales is a biologist with more than 30 years of experience in industry, teaching and environmental research. She holds a PhD of Geography from Birmingham University, School of Geography and Environmental Sciences. Author of many scientific publications and general environmental texts of important subjects for use by students and the non-scientific public.

TABLE OF CONTENTS

GLOSSARY

A

Abiotic – Physical rather than biological.

Abyssal – Relating to or denoting the depths or bed of the ocean.

Agglomerating – Collect or form into a mass or group.

Anoxic – A description of the environment-without oxygen.

Anthropocentric – Regarding humankind as the central or most important element of existence.

Anthropogenic – Chiefly of environmental pollution and pollutants originating in human activity.

Assimilation – The process of taking in and fully understanding information or ideas.

Atmosphere – Is the layer of gases, commonly known as air, that surrounds the planet Earth and is retained by Earth's gravity.

B

Benefit-cost analysis – A method of comparing alternative actions according to the relative costs incurred.

Biodiversity – The variety of plant and animal life in the world or in a particular habitat.

Biosphere – Is the worldwide sum of all ecosystems.

Bog – A bog or bogland is a wetland that accumulates peat, a deposit of dead plant material—often mosses, and in a majority of cases, sphagnum moss.

C

Carbon Footprint – Is historically defined as the total greenhouse gas emissions caused by an individual, event, organization, or product.

Compliance – The action or fact of complying with a wish or command.

Constraints – A limitation or restriction.

Contamination – The action or state of making or being made impure by polluting or poisoning.

Contentious – Causing or likely to cause an argument.

Corrosion – The process of corroding or being corroded.

D

Decomposition – The state or process of rotting; decay.

Denitrification – He loss or removal of nitrogen or nitrogen compounds.

Dispersion – The phenomenon in which the phase velocity of a wave depends on its frequency.

Disseminating – Spread (something, especially information) widely.

E

Ecosystem – Is a community of living organisms in conjunction with the nonliving components of their environment, interacting as a system.

Eel – Any ray-finned fish belonging to the order Anguilliformes, which consists of four suborders, 20 families, 111 genera, and about 800 species.

Emission – The production and discharge of something, especially gas or radiation.

Epidemiologic – Relating to the branch of medicine which deals with the incidence, distribution, and control of diseases.

Eutrophication – Excessive richness of nutrients in a lake or other body of water, frequently due to run-off from the land, which causes a dense growth of plant life.

Excavated – Make (a hole or channel) by digging.

F

Flora – All the plants found in a given area.

Foliage – Plant leaves collectively.

Fragmented – Break or cause to break into fragments.

Fumes – An amount of gas or vapor that smells strongly or is dangerous to inhale.

G

Geostatistics – A class of statistics used to analyze and predict the values associated with spatial or spatiotemporal phenomena.

Green spaces – An area of grass, trees, or other vegetation set apart for recreational or aesthetic purposes in an otherwise urban environment.

H

Horizon – Is the apparent line that separates earth from sky.

Hydrocarbons – A compound of hydrogen and carbon, such as any of those which are the chief components of petroleum and natural gas.

Hydrogeology – The branch of geology concerned with water occurring underground or on the surface of the earth.

Hydrology – Is the scientific study of the movement, distribution, and management of water on Earth and other planets.

Hydrosphere – Is the total amount of water on a planet.

J

Janthina – Is a genus of small to medium-sized pelagic or planktonic sea snails.

L

Lacunae – An unfilled space; a gap.

Larvae – Is a distinct juvenile form many animals undergo before metamorphosis into adults.

Lithology – The study of the general physical characteristics of rocks.

Lithosphere – The solid, outer part of the earth.

M

Meteorology – A branch of the atmospheric sciences which includes atmospheric chemistry and atmospheric physics, with a major focus on weather forecasting.

Mitigation – The action of reducing the severity, seriousness, or painfulness of something.

N

Neritic – Relating to or denoting the shallow part of the sea near a coast and overlying the continental shelf.

O

Obstruction – The action of obstructing or the state of being obstructed.

Organic Matter – The large source of carbon-based compounds found within natural and engineered, terrestrial, and aquatic environments.

Ornithologists – Is a branch of zoology that concerns the study of birds.

Outmoded – Old-fashioned.

P

Peat – Also known as turf, is an accumulation of partially decayed vegetation or organic matter.

Pedogenesis – The process of soil formation as regulated by the effects of place, environment, and history.

Pedology – Another term for soil science.

Perish – Die, especially in a violent or sudden way.

Permafrost – In geology, permafrost is ground, including rock or soil, with a temperature that remains at or below the freezing point of water 0°C for two or more years.

Phytoplankton – Very small plants that float near the surface of water and on which sea creatures feed.

Portability – The ability to be easily carried or moved.

Posterity – All future generations of people.

Precipitation – The action or process of precipitating a substance from a solution.

R

Refurbishment – The renovation and redecoration of something, especially a building.

Retrofitting – Refers to the addition of new technology or features to older systems.

S

Salmon – Is the common name for several species of ray-finned fish in the family Salmonidae.

Scavenging – Search for and collect (anything usable) from discarded waste.

Sedimentation – The process of settling or being deposited as a sediment.

Sludge – Thick, soft, wet mud or a similar viscous mixture of liquid and solid components, especially the product of an industrial or refining process.

Soil – The upper layer of earth in which plants grow.

Soil Erosion – The displacement of the upper layer of soil, it is one form of soil degradation.

Solum – It consists of the surface and subsoil layers that have undergone the same soil-forming conditions.

Syndrome – A group of symptoms which consistently occur together, or a condition characterized by a set of associated symptoms.

T

Taxonomists – A biologist that groups organisms into categories.

Topography – The arrangement of the natural and artificial physical features of an area.

Triple Bottom Line – The triple bottom line is an accounting framework with three parts: social, environmental, and financial.

V

Vellela – Is a monospecific genus of hydrozoa in the Porpitidae family.

Volatile organic compounds – Organic chemicals that have a high vapor pressure at ordinary room temperature.

W

Walkability – A measure of how friendly an area is to walk.

LIST OF FIGURES

LIST OF TABLES

LIST OF ABBREVIATIONS

AADT	annual average daily traffic
AD	anaerobic digestion
AIA	association for impact assessment
AOP	advanced oxidation process
AQGs	air quality guidelines
AQIA	air quality impact assessment
AQMA	air quality management area
ATSDR	Agency for Toxic Substances and Disease Registry
AWB	Asian wetland bureau
BCWQI	British Columbia water quality index
BEC	biological ecosystem classification
BLUE	best linear unbiased estimator
BOD	biochemical oxygen demand
CA	cluster analysis
CAFOs	concentrated feeding operations
CBA	cost-benefit analysis
CCA	canonical correlation analysis
CH_4	methane
CO	carbon monoxide
CO_2	carbon dioxide
CSO	combined sewer overflow
DEAT	Department of Environmental Affairs and Tourism
DEWATS	decentralized wastewater systems
EcE	economic-cum-environmental
EHS	environmental, health, and safety
EIA	environmental impact assessment
EIO-LCA	environmental input-output life cycle analysis
EIS	environmental impact statement

EMS	environmental management system
EPA	environmental protection agency
GRoWI	global review of wetland inventories
HIA	health impact assessment
IRIS	integrated risk information system
ISBL	inside the plant battery limits
ISWM	integrated solid waste management
IT	interim target
IWRB	International Waterfowl and Wetland Research Bureau
LCA	life-cycle assessment
LMW	low molecular weight
MedWet	Mediterranean wetland initiative
MSWM	Municipal Solid Waste Management
NAPPO	Namibia Has An Atmospheric Pollution Prevention Ordinance
NEPA	National Environment Policy Act
NIMBY	not-in-my-backyard
Nj	dinitrogen molecule
NOx	oxides of nitrogen
NPDES	National Pollutant Discharge Elimination System
NPP	net primary production
NPPF	national planning policy framework
NPS	non-point source
NSFWQI	National Sanitation Foundation Water Quality Index
O_3	ozone
OK	ordinary kriging
PAH	polycyclic aromatic hydrocarbons
Pb	lead
PC	principal components
PCA	principal component analysis
PCB	polychlorinated biphenyls
PCs	principal components
POPs	persistent organic pollutants
RZWT	root zone wastewater treatment
SABS	South African Bureau of Standards

SEA	strategic environment assessment
SIA	social impact assessment
STPs	sewage treatment plants
SWAT	soil and water assessment tool
SWM	solid waste management
SWOT	strength, weakness, opportunity, and threat
TBL	triple bottom line
TEEB	The Economics of Ecosystems and Biodiversity
UNCCD	United Nations Convention to Combat Desertification
UNEP	United Nations Environment Program
USEPA	United States Environmental Protection Agency
VEC	valued ecosystem component
VOC	volatile organic compound
WA	wetlands for the Americas
WHO	World Health Organization
WRI	World Resources Institute
WSPs	waste stabilization ponds
WWTFs	wastewater treatment facilities

PREFACE

Environment is a crucial part of the world we live in. It is the basic and most fundamental block of the human life and directly affects the quality of life and even its length of not only the humans but also the other species on this earth.

As the humans have been the ones that have developed and evolved the most of all the species around the world, they have used environment and even gone on to exploit it on many occasions rendering the environment incapable to support the various forms of life in many parts of the world.

It is thus the moral obligation of the humans to look after the environment and modify their actions and plans to suit the conservation of environment and betterment of health around the world. Hence, there needs to be various environmental measures in place that keep a check on the human activities that have an impact on the environment in direct or indirect ways.

The measures and strategies thus formed require a basis to be developed and a rationale on which they may develop. They require an assessment of the various factors and variables that guide the environment and indicate the condition of the environment. This assessment is generally known as the environmental impact assessment.

The environment impact assessment covers a vast range of topics and factors. It focuses on the way in which the solid waste is being treated by the various organizations before it is being released into the environment, and also takes care of the treatment of the various releases and emissions of gases and fluids from the various industries and machines, including the automobiles and the natural emissions.

The assessment can take place in various ways and employs a wide range of tools and methods. The process involves a lot of processes, which make sure that all the aspects of the environment covered in the assessment comprehensively. The assessment also employs a wide range of technologies in the whole process, so as to ensure accuracy and precision in the assessed values.

There are several other areas in which the environmental impact assessment is employed which may be among the land and soil assessment, the various biological components and the assessment of the water bodies, even going on to focus on the effect of noise on the environment. All these areas are covered in the book in-depth and have been explained to the readers with utmost detail.

This book thus tries to bring the issue of environmental degradation to the attention of the readers and inform them about the various steps that are being taken to ensure that the environment problems can be assessed properly.

I hope that the knowledge and information provided in the book help and motivates the readers and various other scholars to find the solutions to those problems and design new solutions to deal with the problems.

CHAPTER 1

Introduction to Environmental Impact Assessment

CONTENTS

The chapter of introduction to environmental impact assessment explains the basic significance of the environmental impact assessment. This chapter also explains the evolution as well as the history of the environmental impact assessment. This chapter mentions about the process of environment such as screening, scoping, impact analysis, mitigation, reporting, review of environmental impact assessment, decision making, and post monitoring. This chapter also includes the various types of impact assessment such as strategic environment assessment, regional environmental impact assessment, and sectoral environmental impact assessment.

This chapter also provides highlights on the basic principles of the EIA such as Proactive Planning and Decision Tool, Avoidance, Pre-emption, and Prevention of Adverse Environmental, Making Positive Influence on Decision Making at the Earliest Possible Opportunity and Thinking Proactively about Options and Alternatives, Living Process Throughout the Project Cycle, Making EIA Recommendations Enforceable, Flexibility Amidst Robustness and Transparency. Along with these principles, there are various other aspects that are discussed in this chapter.

1.1. INTRODUCTION

Environment Impact Assessment or EIA can be defined as the study to predict the impact of a proposed activity or a project on the environment. EIA is a decision-making tool. This tool compares several alternatives for a project. It tries to identify the one alternative that represents the best combination of economic as well as environmental costs and benefits.

Environment Impact Assessment or EIA systematically studies both the beneficial as well as adverse consequences of the project and it makes sure that these effects are considered during the design phase of the project. In addition to it, it also helps to recognize all the possible environmental effects of the proposed project. It also proposes measures to lessen the adverse impacts and predicts whether there will be significant adverse environmental effects, even after the amendment is implemented.

By the consideration of the environmental effects as well as mitigation early in the project planning cycle, environmental assessment has various advantages. Some of these benefits include such as optimum use of resources, protection of environment, and saves overall time and cost of the project.

If EIA is properly conducted, it also reduces conflicts by encouraging community participation, tells decision-makers, and it also helps in laying the base for environmentally sound projects. All the advantages of integrating Environment Impact Assessment or EIA have been observed in all stages of a project, from the stages of exploration and planning, through construction, operations, decommissioning, and beyond site closure.

1.1.1. Objectives of Environmental Impact Assessment (EIA)

The main objectives of the Environmental Impact Assessment are:

- To make sure that Environmental considerations are addressed in a proper way and these considerations should be incorporated into the decision-making process;
- To avoid, reduce or balance the adverse significant biophysical, social as well as other relevant effects of developmental projects;
- To safeguard the productivity and capacity of the natural system and ecological processes in order to maintain their function; and
- To encourage development that is sustainable and optimize the use of resources and all the management opportunities.

1.1.2. Characteristics of Environmental Impact Assessment

These are the following characteristics of an ideal EIA:

- It is applied to all activities that have a significant environmental effect and address all the effects that are expected to be significant;
- Compare alternatives to a proposed project (which includes the possibility of not developing the site), techniques, management, and mitigation measures;
- Clear EIS stating the significance of the impacts along with mentioning their specific characteristics to the experts and the amateurs in the field;
- Participation of the public and strict procedures followed for administrative review;
- Providing the information on the correct time for decision making enforcing the changes; and
- It includes monitoring as well as feedback procedures.

1.2.3. Benefits of EIA

There are several benefits to conduct an Environmental Impact Assessment (EIA). There are several governments as well as donor agencies that make use of EIA in order to understand and execute the complex environmental issues, like waste management or water management techniques.

There are some countries that have specific legal as well as institutional regulations before conducting an EIA. An EIA is a part of the Resource Management Acts in many countries. These countries are particularly interested in preserving as well as maintaining their environments like Canada and Australia.

Some of the general benefits of an EIA comprise:

- reduced time of project implementation;
- cost-saving; and
- adhering of legal regulations.

EIA is economically feasible. This is because it takes comparatively minimal time as compared to other environment assessment techniques. The other techniques might not be precise and might not provide advantages to the organization.

The other benefit of EIA is that if it is followed in accordance with the methods that are involved, it can be used to protect the trees, waste, water, forestry, agriculture, recreational, and cleanliness of cities. These are considered as the important benefits for positive as well as vibrant city planning.

It has been observed that there will also be increased communal activity as well as knowledge regarding all those resources that are included during the assessment of the EIA. For some of the important environmental concerns like climate change, the society, as well as the community, can largely be included in assessing the methods that are used to combat such kinds of challenges.

In countries such as India where water is very sparse and unregulated, with the shortage of water, EIA would be considered as an important technique. This technique is used to measure water and implement techniques like rainwater harvesting in the neighborhoods.

Since it has been observed that the government of many developing countries such as India is not progressive regarding the environmental issues, citizens of that nation are taking the issue of water shortage in their own hands with the help of private companies. Such community involvement is very important and required for the discussion as well as implementation of Environmental Impact Assessment or EIA.

There are several developing countries that fail to use it according to standards and regulations. This is because research and regulations are involved in EIA. It has been observed that private environmental organizations are paving the way forward for the implementation of EIA in the developing world. This is required to be challenged as government agencies become more informed about the techniques of methods that are used in EIA.

1.2.4. Functions of EIA

There are the following benefits of EIA:

- It provided an opportunity for public participation;
- Reduced costs of project and delays;
- The sustainable utilization of natural resources;
- Increased protection of human health;
- Reduced risks of environmental disasters; and
- Increased government responsibility.

The purpose of EIA is to enhance a trade-off between developmental activities and socio-ecological losses. It is a management tool that is associated closely with the project that is providing appropriate environmental information within the stipulated time.

1.2. EVOLUTION AND HISTORY OF EIA

It has been observed that environment Impact Assessment or EIA is one of the successful policy innovations of the 20th Century for environmental conservation.

There was no EIA thirty-seven years ago. But in the present times, it is a formal process in many countries and is presently practiced in more than 100 countries.

> Environment Impact Assessment or EIA as a mandatory regulatory procedure originated in the early period of 1970s. It was originated with the implementation of the National Environment Policy Act (NEPA) 1969 in the United States.

It has been observed that much of the initial development was in a small number of high-income nations such as Australia, Canada, and New Zealand (1973–74). Though, there were also some developing countries Columbia (1974), Philippines (1978), which introduced EIA comparatively early.

The environment Impact Assessment or EIA process really took off after the middle of the 1980s. In the year 1989, the World Bank adopted EIA for some of the major development projects. In this project, the borrower country had to undertake the EIA under the supervision of the Bank.

Table 1.1: Evolution and History of EIA

	Development of EIA
Pre-1970	• Project review that is based on the engineering/ technical as well as economic analysis. • Limited consideration is given to all the environmental consequences.
Early/mid -1970s	• EIA was introduced by NEPA in the year 1970 in the United States. • Basic principle: Guidelines, procedures which include public participation requirement started. • There was the development of standard methodologies for impact analysis (for example, checklist, matrix, and network). • Australia, Canada, and New Zealand became the first countries to follow NEPA in the year 1973–1974. Canada and New Zealand established administrative procedures, unlike Australia, which legislated EIA. • Major public inquires help in shaping the process development.
Late 1970 and early 1980s	• More formalized guidance. • There were other industrial and developing countries that introduced formal EIA requirements (France, in the year 1976; Philippines, in the year 1977). These countries started to use the process informally or experimentally (Netherlands, 1978) or adopted elements, like impact statements or reports, as part of development applications for planning permission (German states [lander], Ireland). • EA was used by developing countries (such as China, Philippines, Brazil, Indonesia) • Strategic Environment Assessment (SEA), risk analysis was involved in the process of EA. • More focus on ecological modeling, prediction as well as evaluation methods. • Coordination of EA with land use planning processes. • Provision for public involvement.
Mid-1980s to end of decade	• EC Directive on EIA establishes basic principle and procedural requirements for all member states, in Europe, • Increasing efforts in order to address all the cumulative effects. • World bank and also, there were other leading international aid agencies that establish EA requirements. • In Asia, there was the spread of EIA process.

1990s	• Requirement to consider trans-boundary effects under Espoo convention. • In this period, there was increased use of GIS and other kinds of information technologies. • Much attention was given to Sustainability principal as well as global issues. • Many developing countries such as India also adopted the EIA formally. • Formulation of EA legislation by many developing countries. • In this period, there was a rapid growth in EA training.

Source: Table from rlarrdc

1.3. THE EIA PROCESSES

Figure 1.1: The process of EIA.

All the stages that are involved in EIA process will depend upon the requirements of the nation or donor. However, it has been observed that most Environment Impact Assessment or EIA processes have a common structure. And, the application of the main stages is a basic standard of good practice.

There are eight steps in the environment impact assessment. Every step in EIA are equally important in determining the overall performance of the project. The Environment Impact Assessment or EIA process usually starts with screening to confirm that the time and resources are directed at the proposals that matter environmentally.

On the other hand, this process ends with some form of follow up on the execution of the decisions and actions taken as a result of an EIA report. The eight steps that are involved in EIA process are explained below:

1.3.1. Screening

This is the first stage of EIA. It determines whether the proposed project need an EIA. And if Environment Impact Assessment or EIA requires EIA, then there is the requirement of level of assessment.

1.3.2. Scoping

This stage of environment Impact Assessment or EIA identifies all the main issues and impact that should be investigated further. In addition to it, this stage also defines the boundary as well as the time limit of the study.

1.3.3. Impact Analysis

This stage of environment Impact Assessment or EIA recognizes and predicts likely environmental and social impact of the proposed project. It also assesses the significance.

1.3.4. Mitigation

This stage in environment Impact Assessment or EIA suggests all the actions that are used to lessen and prevent the potential adverse environmental consequences of development activities.

1.3.5. Reporting

This step-in environment Impact Assessment or EIA presents the result of EIA. The result is presented in a form of a report to the decision-making body and some other interested parties.

1.3.6. Review of EIA

This stage of environment Impact Assessment or EIA examines the adequacy as well as efficiency of the EIA report. In addition to it, it provides information that is required for the process of decision-making.

1.3.7. Decision-making

This step-in environment Impact Assessment or EIA decides whether thc project is approved, or it is rejected, or the project needs further change.

1.3.8. Post Monitoring

Once the project is commissioned, this stage comes into play. It checks whether the effects of the project do not exceed the legal standards. In addition to it, it also checks whether the implementation of the mitigation measures is in the manner as it is described in the EIA report.

1.4. FORMS OF IMPACT ASSESSMENT

Figure 1.2: Different forms of impact assessment.

There are several forms of impact assessment such as health impact assessment (HIA) and social impact assessment (SIA). These are used to measure the health as well as the social consequences of development so that they are considered along with the environmental assessment.

1.4.1. Strategic Environment Assessment

This form of impact assessment that is a strategic environment assessment (SEA) refers to the systematic analysis of the environmental effects of development policies, programs, plans, and other proposed strategic actions.

This process helps in extending the aims as well as principles of EIA upstream in the process of decision-making, beyond the project level and also, when main alternatives are still open. In addition to it, it also represents a hands-on approach for integrating various environmental considerations into higher-level decision-making.

In spite of its wide use and acceptance, EIA also has some shortcomings as a tool for reducing the environmental impacts of development proposals. It usually occurs relatively late at the downstream end of the process of decision-making. It takes place after the major alternatives for the process have been finalized and the flow of the process has been finalized.

1.4.2. Regional EIA

In the context of regional planning, EIA integrates environmental concerns into development planning for a geographic region, generally at the sub-country level. Such a kind of approach is referred to as the economic-cum-environmental (EcE) development planning.

This kind of approach also enables adequate integration of economic development along with the management of renewable natural resources within the carrying capacity limitation. This helps in achieving sustainable development. In addition to it, it also fulfills the need at the macro-level environmental integration. It has been observed that the project-oriented EIA is not able to address it in an effective manner.

The environmental impacts of regional development plans are addressed by regional EIA and thus, the context for project-level EIA of the subsequent projects, within the region. In addition to it, if all the environmental impacts are considered at the regional level, then the cumulative environmental impacts of all the projects inside the region can be accounted.

1.4.3. Sectoral EIA

An EIA should take place in the context of regional as well as sectoral level planning, instead of project-level-EIA. The scope of project-level EIA will be quite minimal. This will be quite minimal once sectoral level development plans have the integrated sectoral environmental concerns addressed. There are several environmental problems that may be encountered in the process of planning and implementing sectoral development projects. Sectoral EIA will help in addressing these specific environmental problems.

It has been observed that SEA that is SEA had limited development and implementation till the year 1990. Although, after the year 1990, several countries in developed economies adopted SEA. Some of the nations like Denmark and Canada have made provision for SEA of plans, policy, and programmes. These countries made it separately from EIA legislation and procedure.

Table 1.2: Difference in EIA and SEA

Environment impact assessment	Strategic environment assessment
• It takes place at the end of the decision-making cycle. • There is a reactive approach to the development proposal. • It recognizes specific influences on the environment. • A limited number of feasible alternatives are considered. • A limited review of cumulative effects. • Focus on mitigating as well as minimizing impacts. • Environment impact assessment has a narrow perspective along with a high level of detail. • It is a well-defined process. It has a clear beginning as well as the end. • Emphasis on standard agenda, treats the symptoms of environmental degradation	• It takes place at earlier stages of decision-making cycle. • There is a pro-active approach to development proposals. • It also identifies environmental implications, problems of sustainable development. • Broad range of potential alternatives are considered. • Early warning of cumulative effects. • Focus on meeting environmental objectives, also, on maintaining natural systems. • It has a broad perspective, lower level of detail in order to provide a vision and overall framework. • It is a multi-stage process, policy level is continuing, iterative, overlapping components. • Emphasis on sustainability agenda, approaches the major sources of environmental deterioration.

Source: Source: Table from rlarrdc

There were several other countries such as Slovakia, Czech Republic, etc., which have introduced SEA requirements through reforms in EIA legislation. On the other hand, in the case of the United Kingdom, SEA requirements were introduced through environmental appraisal.

It is a part of resource management or biodiversity conservation regimes in some countries like New Zealand and Australia. It has been observed that the adoption of SEA is likely to grow significantly in the upcoming years particularly with directives by the European Union and Protocol to the UNECE Convention on Transboundary EIA by signatory countries (with a provisional date of May 2003 for completion).

Table 1.3: Comparative Review of EIA Procedures and Practices

Developed countries	**EIA in developing countries**
In developed countries, there is a well-framed EIA legislation in place. For example, in Canada, the Canadian Environmental Assessment Act regulates EIA. On the other hand, EU countries are guided by Directive on EIA (1985).	There is the absence of formal EIA legislation in many of the developing countries. For example, EIA is not mandatory in many of the African countries.
There is the active involvement of all participants in the developed countries, it includes competent authority, government agencies and affected people at the early stages of the EIA. This helps in making the process more robust and it also provided a fare idea of issues, which are required to be addressed at the initial phase of EIA.	In developing countries, there is limited involvement of public and government agencies in the initial phases of EIA. Often, this results in poor representation of the issues and impacts in the EIA report. It adversely effects the quality of the report.
In developed countries, integrated approach to EIA followed. All aspects which includes social as well as health are considered.	In developing countries, mainly environmental aspects are taken into account. Poor on social or health aspects.
In developed countries, there is a proper consideration of alternatives in EIA	In developing countries, the consideration of alternatives is more or less absent.
There is a well-defined process of screening. For example, in EU countries competent authorities decide whether EIA is needed after looking for advice from developer, NGO, and statutory consultees. For instance, In Japan, screening decision is made by the authorizing agency. This decision is made with respect to certain criteria. On the other hand, in Canada, federal authority determines whether an environmental assessment is needed or not.	Screening practice in EIA is weak, in developing countries. In most of the cases, there is a list of activities that need EIA but without any threshold values.

In developed countries, scoping process is comprehensive, and it also includes consultation with all the stakeholders. For example, in many countries such as the United States, Netherlands, Canada, and Europe, the involvement of public and their issues are addressed in scoping exercise. Besides this, there are funding organization like ADB, World Bank, and ERDB have provision to consult the affected people and it also have NGOs during identification of issues in scoping exercise.	In most of the developing countries, the scoping process is very poorly defined. In many countries including China, Pakistan, etc. there is no provision for scoping. In some of the countries such as in Nigeria and Indonesia, a term of reference is followed for scoping. On the other hand, in some other countries such as Chile, Ghana, and Taiwan, a general checklist is followed. There is no public consultation during scoping in the countries where it is undertaken, In addition to it, in most of the developing countries, scoping is often directed towards meeting pollution control requirements, rather than addressing the full range of potential environmental effects from a proposed development.
In developed countries, most of the reports are in local language.	In developing countries, most of the reports are in English and not in the local language.
A multi-disciplinary approach. There is the involvement of expert with expertise in various areas.	There is the absence of trained EIA professional. This often leads to the preparation of insufficient and irrelevant EIA reports in developing countries
In developed countries, there is two-tier of EIA review, one is conducted after the completion of EIA. This is done to check the adequacy and effectiveness of EIA and the second one is done before the decision-making.	In developing countries, there is poor review or monitoring.
Expertise in EIA: The International Association for Impact Assessment (AIA) and all the other organizations show that there are a large number of people with the capability to design, conduct, review, and evaluate EIAs from all the countries of the North. In addition to it, the main portion of teaching about environmental assessment also takes place in industrial countries.	In developing countries, the expertise in EIA is slowly developing. In most of the cases, students from the developing countries go to the developed countries in order to gather knowledge of the subject.

Source: Table from rlarrdc.

1.5. BASIC PRINCIPLES OF EIA

1. Principle One: Proactive Planning and Decision Tool

The Environment Impact Assessment or EIA process is a proactive planning tool. This tool is used to avoid and pre-empt adverse environmental impacts. These impacts might be caused or experienced by the proposed development. This EIA process should be associated to the decision-making process.

2. Principle Two: Avoidance, Pre-emption, and Prevention of Adverse Environmental

The consequences of the main objective of the EIA process are to prevent and avoid adverse environmental consequences of the proposed project. If in case the adverse environmental impacts cannot be completely avoided, measures should be taken into account in order to reduce and control the possible adverse environmental impacts within the pre-defined and strict limits or criteria.

3. Principle Three: Making Positive Influence on Decision Making at the Earliest Possible Opportunity and Thinking Proactively About Options and Alternatives

It is very important to facilitate any effort to improve the environmental performance of the proposed development at the initial phase of the project planning instead at the later stage of the project development.

Often, it is better to pre-empt all the adverse impacts at the initial stage of the project implementation rather than the later stages. When adverse environmental impacts are expected, options, as well as alternatives, should be taken into account, and in considering alternatives, the main consideration is whether applying the precautionary principle, the proposed project is presented to be within the environmental criteria and is environmentally acceptable.

4. Principle Four: Living Process Throughout the Project Cycle

The EIA process should include a course of dynamic actions throughout the cycle of the project to identify and predict all the possible adverse environmental impacts from the proposed development, to develop all that measures which are used to avoid, pre-empt, prevent, and decrease the environmental problems, and to manage as well as control environmental impacts to within established limits or criteria.

5. Principle Five: Making EIA Recommendations Enforceable

All the EIA recommendations or the suggestions should be practical, sensible, and effective, with information about the 5 Ws (that is what are the mitigation measures that would be implemented, by whom, when, where, and to what requirements).

These recommendations should also have the clear definition of the responsibility for implementing all the recommended mitigation measures. The recommended measures should be easy to enforce and can avoid all the environmental issues from occurring, rather than depending on remedial measures after some issues or problems happen.

6. Principle Six: Flexibility Amidst Robustness and Transparency, With Public Participation and With the Ability to Adapt to Changes

The EIA process should be transparent, and it should also consider public participation, but this process should be flexible to adapt to all the altering circumstances without compromising the environmental requirements.

7. Principle Seven: Seeking Practical Environmental Outcomes for the Environment and Community

To deliver and communicate practical environmental outcomes for the environment and also for the community is one of the purposes of the EIA process.

8. Principle Eight: Avoidance of Any Late Focus

The early emphasis on the main adverse environmental consequences could help in saving a large amount of efforts or costs. This may otherwise increase because of the expensive or time-consuming remedial works at the later stage.

9. Principle Nine: Efficiency Amidst Effectiveness

An effective and efficient approach in carrying out EIA studies step by step would enhance the productivity of the EIA process.

10. Principle Ten: Transparent Agreement Among Relevant Parties, Clear Expectations of What Need to Be Done and What the Performance Will Be, and Explicit Resolution of Any Conflicts

The expectations, agreements, performance requirements and any conflict resolutions, etc. should be completely communicated among all the relevant parties. It should be communicated in an open and frank manner in order to avoid misunderstanding and also, to improve better management of EIA follow up works.

Table 1.4: Valued Ecosystems Components

Environmental Resource	Valued Ecosystem Component	Importance of Valued Ecosystem Component
Air and Climate	Air quality	Effects on air for all the residents of the local area. Health implications for all the users. It also effects ecosystem
	Climate	It also contributes to global warming.
Land	Geomorphology and Landscape	Application of non-renewable resources. Great importance to local community. Effects of waste disposal methods. Changes in land morphology.
Water	Surface Water Quality	Sustainability issues. Has an impact on local usage of water (such as for the purpose of irrigation, drinking, and transportation) Health implications for all the users
	Groundwater Quality	Sustainability issues (that is, is groundwater considered to be a highly limited renewable resource in the region?) Hasan impact on the local use (that is irrigation and drinking).
	River Water Quality	Has an impact on the local use (transportation, fisheries) Health implications for all users

Ecology and Biodiversity	Marine Ecology and Biodiversity	It is of great importance for the wellbeing of all biological components of the ecosystem. High biological value (that is International, National, and Regional) Economic use to the society
	Terrestrial Ecology and Biodiversity	It is essential for ecosystem wellbeing and also, for the proper functioning. Importance to biodiversity value. (that is International, National, and Regional) Usefulness in the community
Human Environment	Community Health and Safety	Operations impact on the safety of the community Decrease in the rate of gas flaring
	Socio-Economic Activities	It is important for employment opportunities Community welfare
	Light Pollution	Annoyance to the local community and also, to the ecosystem
	Noise Pollution	Impact on the biological diversity Nuisance to local community
	Agriculture	Socio-economic importance National and community value creation

Source: Table from opic.gov

1.6. DIFFERENT COMPONENTS AND ENVIRONMENTAL IMPACT ASSESSMENT

1.6.1. Air and Climate

Figure 1.3: Air and climate are heavily impacted by the pollutants released during various industrial activities.

Source: Image by PxHere.

The quality of the air may be affected during site preparation, any construction activities, and it can also be affected by the potential occurrence of non-routine events. Different kinds of pollutants and their impacts are discussed below.

Dust and Particulates

Figure 1.4: Dust and Particulates are responsible for heavily contaminating the climate.

Source: Image by Wikipedia.

At the time of construction, there may be a localized and temporary reduction in the quality of the air as a result of dust and particulate generation. It is significant, with the potential to affect all the workers on-site. Although, the impact on off-site receptors will be negligible as the actual Project site is circa that is 7 kilometers from the nearby habitation.

The creation of excavation, access roads, and earthworks ISBL (inside the plant battery limits) may give rise to the temporary airborne transport of particulates (that is the increased dust). The majority of the dust that is generated at the time of construction is possibly to be deposited within 100 meters of the site.

According to the United Kingdom Environment Agency Technical Guidance Document (Monitoring) M17, it was reported that almost 85% of the particulate matter in the ambient air is deposited within the 100 meters of the source. On the other hand, only 10% deposition was reported between the distance of 100 and 500 meters from the source.

It is very important to note that the whole area that is around the Project site is basically undeveloped for over 3 kilometers radius around the site. Indeed, it has been observed that most of the undeveloped land that is around the site is owned by the Project Sponsor. The nearby residential area is situated about 7 kilometers from the project site.

The quality of the air will also be reduced by releases from machinery and vehicles that are used for transportation to and from the project site and on the site. These releases are not anticipated to be significant.

Gaseous and Exhaust Emissions

There are the emissions of pollutants at the time of testing and start-up phase of the Project. The gaseous emissions may consist of CO_2, NO_x, NH_3, CO, SO_x, and particulate matter. The start-up phase of the project is needed to enable testing of the machinery and installations. The start-up/pre-commissioning stage of the project usually lasts for few weeks.

In addition to it, heavy equipment like bulldozers and other kinds of construction equipment will produce exhaust emissions from diesel engines. This gives rise to temporary increase in the concentrations of NO_x, SO_x, and CO_2. Emissions of these gases may occur in the event of an emergency. The major gases of concern consist of:

1. Sulfur Dioxide – SO_2

The amount of SO_2 that is present in the exhaust gases is directly dependent on the content of sulfur of the used fuel. The reduction of the emissions of SO_2 from engines can be implemented by the use of low sulfur fuel for machinery as well as equipment. The Project Company will tell all the contractors to make use of low sulfur fuel, to the extent that is available locally.

2. Nitrogen Oxides – NOx

The emission of NO_x from contractor equipment/activities will require to be minimized. Also, the construction equipment and machinery are required to be modified, to the extent that is required.

3. Carbon Oxides

Carbon dioxide (CO_2) is a greenhouse gas and emissions may take place during the construction phase. The emission of carbon dioxide is not directly harmful to the health of human or the environment. But this gas contributes to global warming as well as climate change.

During the construction phase of the project, carbon dioxide may be released and generated both directly and indirectly.

It is very important to note that all energy-consuming, and CO_2-generating activities are carried out as efficiently as possible in order to minimize the emission of CO_2. The incomplete combustion of fuel in engines results in the emission of carbon monoxide (CO). Under normal conditions, it has been observed that air emissions will have minor impacts on the surrounding environment at the time of construction phase.

1.6.2. Land

Construction activities may have impact on the land during site preparation. It is essential to note that the Project area is benign when considered from a geological perspective. The project area generally has no known history of tornadoes, flooding, earthquakes or other climatic nor seismic events.

However, site clearance and ground leveling, which is required for plant construction, will change the geomorphology of the site. It has been observed that the area which is designated for the construction of the pipeline, process plants, and product storage are not considered as areas where geological features need protection.

It has to be noted that earthworks and excavation are not regarded as sources of negative impact on local geology. As a means of recovery, surface deposits that will be eradicated at the time of construction phase may be used for the purpose of leveling or backfill. This is done to raise and restore the landscape. Generally, it has been observed that no significant impacts on geology are considered likely at the time of the project construction.

The site area will be re-landscaped during the construction phase of the Project. Any existing vegetation will be eradicated, the soil will be leveled, graded, and compacted in order to produce a flat and even surface.

The Plant facilities, as well as installations that are completed during the construction phase, will be visible and these may also have a negative visual impact. Although, it has been observed that this is consistent with the designation and licensing of the site for the development of a petrochemical complex. In addition to it, there are already several numbers of gas and oil installations and infrastructure in the Project site.

Solid waste that is generated during the construction phase may also have a negative impact on the project site if it is handled in an inappropriate way, but there are contractors who will establish a (SWM) plan. This plan is developed in order to ensure that any solid wastes that is generated by the construction activities are handled in an appropriate manner.

1.6.3. Water Groundwater Quality

Groundwater is a major VEC that is required to be preserved and monitored at the time of construction of the project. The local community particularly makes use of underground water for subsistence and several other agricultural activities. The project does not expect to make use of underground water. The main plan to pay attention on taking water from the nearby Ossiomo River that is situated beside the site for the Project freshwater intake.

The quality of the groundwater may be affected by construction activities. It is also affected by the occurrence of non-routine events. The main accidental event risk is particularly attributable to seepage or outflow of contaminants from accumulation of solid wastes or improper sanitation facility.

There is a need of large number of personnel (up to 1,000 at the time of peak construction) to work on the project site. The presence of such a large number of workers on the project site will need a sanitation facility which, if not adequate, could give rise to pollution of the surface water, soil, and ultimately groundwater along with the marine water through the introduction of coli form contamination and/or biodegradable organic matter. As a result of this, there is a high biochemical oxygen demand ("BOD") load in certain circumstances. The BOD is defined as a measure of water quality.

As BOD that is BOD load increases, consequently the quality of the water decreases (i.e., water degradation is related with an increase in BOD load).

The quality of groundwater may also be affected by improper waste disposal and spillages at the time of construction. It has been observed that the inadequate management of construction materials and fuel could result into spillages, particularly of machine oil. There are several other hazardous substances that may be used at the time of construction involves solvents, acids, paints, and bases. A spill prevention, as well as a response plan, should be completed in order to prevent any effects to the quality of groundwater.

1.6.3.1. Freshwater Quality

Freshwater that is required for the operation activities will be supplied from the Ossiomo River that is situated beside the Project site. The water that flow on the Ossiomo river at the Project site is approximately (80–120 liters/second) on average. This is based on data that is collected by the Benin-Owena River Basin Authority over the past 30 years.

This is far in excess of the Project requirement during the construction and operation phase. In essence, the use of water from the Ossiomo River in the Project will have no or very little effect on the other users. However, the Project Company and its Sponsors will prioritize to safeguard and protect the existing freshwater users. Also, the Project Company and Sponsors ensure that the increase in the demand from Plant use is within the sustainable limits.

1.7. IMPORTANCE OF ENVIRONMENTAL IMPACT ASSESSMENT

Environmental assessment is taken up as a quick assessment technique that is used for determining the current status of the environment and also, it helps in identifying the actual impact of critical activities on environmental parameters. EIA is a com new planning and decision-making tool that is first enshrined in the United States in the National Environmental Policy Act of 1969.

It is considered as a formal study process that is used to calculate the environmental consequences of any kind of development project. Therefore, EIA make sure that all the potential problems are anticipated and addressed at the initial stage in project planning and design.

The formation of EIA system is very important to conform socio-economic development projects to environmental security and thus it ensures sustainable economic development. Since, development is considered as an ever-growing process, the impact of development on the environment is also ever increasing. This give rise to rapid deterioration in environmental conditions.

As such, a rational approach to sustainable development is provided by environmental assessment provides. In addition to it, it also enables an individual to carry out the environmental cost-benefit analysis of projects. This is carried out in the early stage. Therefore, it is precursor to the complete assessment of environmental impacts, which are taken up only if a requirement for the same is established.

Generally, comprehensive EIA is carried out after the rapid EIA. Sometimes, it is conducted after accruing a formal approval. It helps planning and management in taking long-term measures. These measures are used for efficient management as well as environment conservation.

It has been observed that the environmental variable when establishing a project is becoming more relevant day by day. In the present times, hospital, a new port, or shopping center is not conceived without doing a prior analysis of the consequences that the implementation of the project can bear on the area where it is established physically.

EIA is considered as the most effective method. This is used for preventing aggressions towards environment and also, helps in preserving natural resources in project management.

The Council Directive 85/337/EEC of June 27th, in the year 1985, on the valuation of the effects of certain public and private projects on the environment, takes under advisement that the impacts on a certain project have to be calculated in order to protect the health of the human. This contributes to the improvement of quality life and also, helps in maintaining biodiversity and the conservation of ecosystems.

An EIA shall include a general description of the project, considering environmental impacts, a statement of the major alternatives under study and the data that is needed to recognize and calculate the main effects that the project has on the environment. In addition to it, it shall also mention a description of the measures that are envisioned. These are mentioned in order to avoid, decrease, and, if possible, find the mode of prevention for the adverse effects.

It has to be noted that the assessment always refers to a particular project and it shall describe the different kind of work, the requirements, the materials, and maintenance works at the time of the construction and operational phases, the implemented technologies, etc.

There are a number of measures that can be taken in the sphere of business just to make the project responsible towards the environment after the initiation of the project.

There are several firms such as Garrigues Medio Ambiente, Aguirre Newman and Aire Limpio that presented a report on the environmental situation and sustainability policies of the main Spanish companies in all sectors.

It has been observed that 99% of Spanish companies are concerned about recycle waste, 85% of Spanish big companies are concerned about environment, and 90% monitor the electric consumption. In addition to it, this collected data also reveal that companies are concerned about the consumption of energy and have mainly contributed to regulate ambient temperature. Also, these companies make use of efficient materials and equipment. These materials and equipment include ventilation systems, automatic doors and solar panels.

However, it is not only the big companies that are taking part in this struggle. There are various SMEs that are also the part of this struggle and make helpful actions. It has been observed that between 75% and 90% of them, wrongly think that their activities have no environmental impact.

What can companies do to slow down the climatic change?

Companies can do the following things to slow down the climatic change:

- To be more efficient regarding the consumption of energy;
- To renounce to fossil fuels in order to decrease the rate of emissions;
- Not to leave appliances in standby mode;
- To consume less and also, purchase "green" products;

- To introduce the latest infrastructures and bring up their regular updates without wasting energy;
- To locate workplaces close to homes of workers in order to lower down the daily commuting time; and
- To make use of biofuels.

1.7.1. Legislation

It has been observed that in the present times, there are many countries that have mandatory legislation or policies for EIAs. In some developing countries such as India EIA was formally introduced in the year 1994. Both central, as well as state authorities collectively, share the responsibility of its development and management. There is a post-project evaluation analysis report that is drawn on Strength, Weakness, Opportunity, and Threat (SWOT).

This outlines several constraints, like ineffective monitoring, inappropriate screening, and scope, and provides required guidelines. In addition to it, the legislation also provides realization of opportunity for increasing the awareness among the public, initiatives of several environmental groups and business community and to incorporate environmental consideration into plans as well as policies. There are some of the unexpected threats to the system. These threats are rapid economic reforms, poor governance, and favors to all the small-scale units.

1.8. EIA TECHNIQUES AND METHODS

EIA methods can differ according to how it is used. These methods can range from simple to complicated. This depends on the kind of data that is used and the level of expertise that is needed for the research. There are several qualitative and quantitative including mixed methods that are used to examine an EIA.

The data that is collected is considered as the most important asset. This data will determine the impact of Environmental Impact Assessment or EIA. There are several things such as community involvement, manpower, time, and budget constraints should also be taken into consideration before EIA is done.

EIA will involve much time and effort from the organization in order to gather the required data and implement the technique. This is because EIA is a research-based method. Some methods consist of: Impact Measurement,

Impact Identification, Impact Assessment, and Communication. These criteria are required to be assessed further in order to provide an accurate measurement of data for the EIA.

1.9. IMPORTANCE OF AN EIA IN ENVIRONMENTAL PROTECTION

The project may have an impact on the environment when a project is proposed and designed. These effects include the effect on the quality of water or air, an economic interruption for a community, or even it has an impact on social interactions. These possibilities are required to be evaluated so that all the negative effects can be reduced.

EIAs are the documents that include detailed analyses of proposed projects that may have social, economic, and environmental impacts. They are necessary for main Federal actions and there are some states that need an EIA or a similar document for projects, those projects which can be even more stringent.

The importance of an EIA is to make sure that all the impacts are assessed in order to save the environment and the quality of life for all the human beings and organisms in that vicinity and beyond.

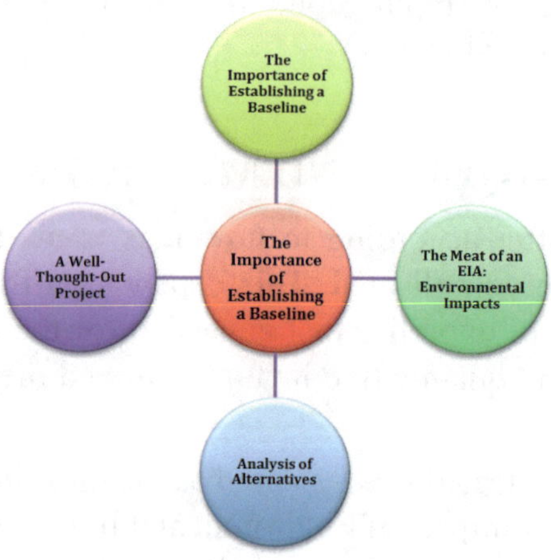

Figure 1.5: The importance of developing a baseline for environmental impact assessment.

1.9.1. The Importance of Establishing a Baseline

The collection and analysis of baseline data is one of the most important components of an EIA. As baseline data exists prior to the project and its potential affects, it is acknowledged as the state of the ecosystem. In an EIA, baseline data explains about the current biological, socioeconomic, and physical conditions of the immediate area and any area that is predicted to be affected.

Scientists collect the Relevant data and the baseline condition is established. The importance of a baseline survey is that it permits the experts or scientists to see how actions are carried out, during the construction project, and impact of these actions on the conditions.

1.9.2. The Meat of an EIA: Environmental Impacts

It has been observed that the EIA explores both positive as well as negative impacts. In addition to it, it even explores possibilities for improving the environment, like enhancing wetland areas as part of the project and improving economic opportunities in the area.

It also recognizes the gaps that are present in baseline data and highlights the uncertainties in predictions. But most importantly, it identifies all the possible negative impacts and suggests some ways that can be helpful in reducing these negative impacts in cases where these are unavoidable.

One of these ways to reduce the negative impacts is to enhance other areas in order to compensate for all the negative impacts in one area. For example, improving native vegetation in those regions that has high surface water run-off. This could help in improving the quality of water and offset all the negative impact.

1.9.3. Analysis of Alternatives

All the reasonable alternatives are taken into consideration, which includes the impact of taking no action. Alternatives also consist of alternative sites, project redesign, and alternative technologies and construction techniques.

The Environmental Impact Assessment or EIA analyzes all of these alternatives with an eye towards reducing the negative impacts to the environment. Also, it permits all the decision makers to select the best alternatives to protect and improve the quality of the environment.

1.9.4. A Well-Thought-Out Project

The EIA permits all the industry project managers, government officials and citizens to take part in the process of developing the EIA. Public comments are considered very important to the process and are sought after in a number of open meetings.

In addition to it, it is required by experts or scientists to respond to various comments and use them into the analysis. The EIA is a very important document that is comprehensive and well thought out. It is also very important to reduce all the negative impacts on the environment.

1.10. NEED OF ENVIRONMENTAL IMPACT ASSESSMENT?

An EIA is defined as an assessment of the impact of a potential project on an environment. This takes into account all the positive and negative impacts, short-, and long-term impacts of the specific actions that will be performed on the particular area.

The assessment is meant to efficiently manage the ecosystems in play; be they social, biophysical or even cultural. There are several advantages to an EIA mainly that once one can predict the impact the project may have; one can take some measures in order to mitigate the effect.

Generally, an EIA is mandatory by the government. There are several international laws that need prevention of damage to certain sites and lessening of influences. These laws will be stricter. This depends on the importance as well as fragility of the ecosystems on which the project will be functioning.

All the methods of the EIA are reliant on the industry and also, there is the need of two follow up practices. The first is a scientific analysis, this is done to analyze the accuracy of the predictions. In addition to it, this will also evaluate and attempt to explain any kind of mistakes or errors that are found in the data and analysis. Management is the second practice that is required. This is where a firm generally comes in to provide assistance to the project with successfully dampening the potential impacts.

1.11. CONCLUSION

EIA is the modern-day method for regulating the use of natural resources. In the modern world, it is very important to understand that the over usage of the natural resources have led to the depletion of the resources. As the industrial era dawned on the world, the natural aspect began to withdraw from the scenario.

Thus, there was a need to assess the role of natural resources in the human activities. Especially when it comes to disposing the waste material from the various industries, the natural resources were heavily exploited. With the degrading quality of the air, water, and land, the EIA became very helpful in making the regulations for waste disposal.

With the advancement of time and development of technology, the methods of Environment Impact Assessment have been upgraded. Also, there has been a higher level of awareness amongst the humankind regarding the usage of natural resources. This has been accomplished with the stringent policies of EIA.

REFERENCES

1. Environmental Impact Assessment Ordinance: Basic Principles of the Environmental Impact Assessment Process, (2010). [eBook] Available at: https://www.epd.gov.hk/eia/hb/materials/GN1.pdf (accessed on 4 March 2020).
2. Environmental Impact Assessment, (2011). [eBook] *Ossiomo AmmoniaUrea Project: Environmental Social Impact Assessment.* Available at: https://www3.opic.gov/environment/eia/greenfields/Chapter%207%20-%20Environmental%20Impact%20Assessment.pdf (accessed on 4 March 2020).
3. Haseeb, J., (2017). *What is Environmental Impact Assessment | Objectives of EIA*. [online] Aboutcivil.org. Available at: https://www.aboutcivil.org/environmental-impact-assessment (accessed on 4 March 2020).
4. Ingole, B., (2015). *Importance of Environmental Impact Assessment and Monitoring Studies in Industrial Development*. [online] Drs.nio.org. Available at: http://drs.nio.org/drs/handle/2264/778 (accessed on 4 March 2020).
5. Introduction to Environment Impact Assessment, (2020). [eBook] *Industry & Environment Unit*. Available at: http://www.rlarrdc.org.in/images/Introduction%20EIA.pdf (accessed on 4 March 2020).
6. Jagannath, T., (2019). *The Importance of Environmental Impact Assessment (EIA) | Planning Tank®*. [online] Planning Tank®. Available at: https://planningtank.com/environment/importance-environmental-impact-assessment-eia (accessed on 4 March 2020).
7. Manusa, (2015). *The Importance of Environmental Impact Assessment-Blog | Manusa.* [online] Blog | Manusa. Available at: http://www.manusablog.com/en/miscelaneus-en/the-importance-of-environmental-impact-assessment/ (accessed on 4 March 2020).
8. Ramchandani, C P., (2010). The Role and Importance of EIA. [online] Ezine Articles. Available at: https://ezinearticles.com/?The-Role-and-Importance-of-EIA&id=4685618 (accessed on 4 March 2020).
9. Scribd, (2020). *Types of EIA | Environmental Impact Assessment |* Environmental Social Science Concepts. [online] Scribd. Available at: https://www.scribd.com/document/246537391/Types-of-EIA (accessed on 4 March 2020).

10. TECSOL Tenement & Environment Compliance, (2020). *Environmental Impact Assessments – EIA Are Mandatory – TECSOL Australia*. [online] Tecsolaustralia.com.au. Available at: http://www.tecsolaustralia.com.au/Why-do-you-need-an-Environmental-Impact-Assessment-bgp2959.html (accessed on 4 March 2020).
11. Terri, S., (2018). *Importance of an EIA in Environmental Protection*. [online] Sciencing.com. Available at: https://sciencing.com/importance-eia-environmental-protection-15523.html (accessed on 4 March 2020).

CHAPTER 2

Methods of Measurement of Environmental Impacts

CONTENTS

The environment impact assessment can be evaluated with the use of measurement-based tools and valuation-based tools. In this chapter, the main focus is on the explanation of the 15 measurement tools and 5 valuation-based tool that measures the impact of environment. The first half of the chapter gives insight about the most common tools that are used in evaluating the environmental impacts. Their uses, advantages, and disadvantages are also explained in some of the measurement-based tools in the mid of the chapter. At the end of the chapter, the valuation tools are discussed.

2.1. INTRODUCTION

As per the report generated by the United Nations in the year 2008, the human activity led to \$6.6 trillion in global environmental damage. Of which, 33% was caused by 3,000 of the world's largest publicly traded organizations (United Nations Environment Program Finance Initiative, 2010).

Today, companies are interested in maintaining a record of their positive and negative environmental impacts. This is required by the companies for setting up their goals more appropriately, for increasing and decreasing their level of negative impacts and for increasing and maintaining their positive impact.

The impact of an environment can be either recorded in measurement or a value. The technique used in the determination of the magnitude of a quantity is a measurement. For instance, an organization can measure the amount of carbon dioxide released annually from a particular chimney. On the other hand, valuing means a monetary value assigned to a particular impact.

However, these terms are not mutually exclusive. Because there are several measurements that are financially based like the profits from the sale of a product. While, there can be some valuations that require some measurements such as in the case of the amount of carbon dioxide from a chimney, before the assignment of a value.

There are numerous tools and measurement techniques that has been designed in order to determine the impact of the product, firm, organization on the environment. However, it is not clear as to which tool is the best for a particular situation.

Although several tools have been studied by the academics, for example, ecological, and carbon footprints, life cycle analysis and ecosystem service valuation. However, there are still some methods which are left to assumptions and interpretation and no consensus exists within individual methodologies.

For example, the owners of an environmental conscious firm who want to minimize the environmental impact they impose while their three employees fly from Toronto to Auckland for a business trip. There is one method to calculate the impact is to calculate the carbon footprint and pay for the carbon offsetting (Gossling et al., 2007; Thomassin, 2003). This would require the search for a carbon offset service, providing the number of the people who are traveling, the destination, mode of travel, etc.

The other method includes the ecosystem service valuation (Costanza et al., 1997; Kaval, 2010; Sandhu et al., 2008). The organization have to decide on the type of method they want to use for the assessment of the environment.

2.2. MOST COMMON TOOLS

According to the database studies, there are total 20 methods or tools, out of which 15 methods are based on measurement methods and other 5 are based on the valuation method. Although these are divided into two categories measurement and valuation, the measurement tools include some valuations and the valuation tools include some measurements.

There are many measurement methods that are based on finance, and many valuations are measurement-based. The lifecycle analysis method is the most widely used method worldwide. It has been found in 180 studies that the lifecycle method is used widely. The other tool that is used widely is the ecological footprint, which is followed by balanced scorecard approach and then the environmental management system (EMS) modeling.

The above-discussed four methods are the measurement methods. The other two methods are valuation-based tools that is most commonly used, i.e., environmental input-output model and ecosystem service valuation. Other tools are less in use.

The studies revealed that the academic used mostly the life cycle analysis method. While, the other three methods were equally popular for the practitioners with the highest percentage use in ecosystem service valuation and then followed by the life cycle analysis and the ecological footprint.

There were several studies that was carried out for finding a new method or updating the old technique with a new name. Among the academics and practitioners, the use of new and revised methods may imply a lack of satisfaction with the currently available methods.

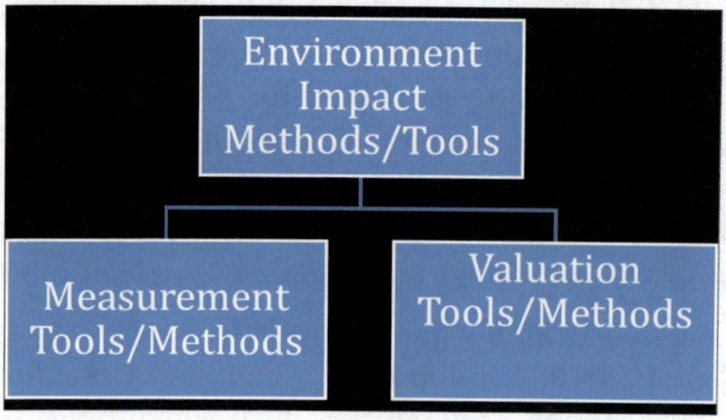

Figure 2.1: Tools for environmental impact assessment.

The below given are the measurement-based methods or tools.

2.2.1. Balanced Scorecard Approach

The tool that is used for recording and evaluating financial and non-financial measures. The Balanced Scorecard Approach is a strategic planning and management tool. Initially, the balanced scorecard was found by Kaplan and Norton (1992), but several other researchers tries to make the approach more balanced by including social, environmental, and financial aspects.

There were other researchers who felt that the approach has all the aspects included in it. By including the triple bottom line (TBL) into a balanced scorecard is the way of building on the original model. There can be addition of the principle environmental impacts.

2.2.2. Carbon Footprint/Carbon Performance Indicators

The calculation of the total amount of greenhouse gas emissions linked to an organization, firm, product or person is the carbon footprint. It is expressed in the form of carbon dioxide. There are some companies that tries to offset their carbon footprint by investing in alternatives such as solar energy, wind energy and reforestation projects (Chambers et al., 2007; Wiedmann and Minx, 2008). This tool or measure is a subset of the ecological, or environmental, footprint, and the life cycle analysis method.

2.2.3. Cost-Benefit Analysis

Cost-benefit analysis is also known as benefit-cost analysis, is an economic tool. For a variety of scenarios, the total expected benefits against the total expected costs is weighed. This tool is used when assessing the case for a new policy proposal, project, product or program. While calculating the complete cost-benefit analysis, total cost associated with social and environment are considered. Thus, it helps to identify the trade-off between environmental and social quality and economic goods (Hussen, 2000).

2.2.4. Destination Environmental Scorecard

A destination environmental scorecard is also called the local environmental activity scorecard. The measurement tool is used by small and medium-sized hotels. This tool measure and compare their performance such that their operations can be made responsible and that benefit both their financial performance and the region's sustainability (Karatzoglou and Spilanis, 2010).

2.2.5. Ecological Footprint

The ecological footprint method measures the human demand on the ability of the affected ecosystems to regenerate the resources consumed and the waste generated. The ecological footprint is also called environmental footprint. Typically, it is measured by the amount of biologically productive land and water area required (Dietz et al., 2007; York et al., 2003).

The ecological footprint tool is also called the environmental footprint and the eco-footprint. This tool is a physical measurement indicator. This tool measures the ecologically productive land and sea required to support human resource demands, such as food, energy, and housing in addition to the assimilation of wastes.

In the year 1990, the concept and calculation method of the ecological footprint as created by the professor William Rees and PhD student Mathis Wackernagel. It was created at The University of British Columbia in Vancouver, Canada (Wackernagel, 1994).

The typical measurement of the environmental impact by the ecological footprint method is in hectares or acres. It can also be calculated for a person, thing, local, regional, or a global area.

In order to be sustainable, annually a person should ideally use one global hectare or less. There has been much advancement and improvements in the use of the ecological footprint method since the seminal method by Wackernagel (1994).

There is an advancement in the creation of Footprint 2.0 by Redefining Progress which is a US-based non-profit organization. When Footprint 2.0 came to function, the Wackernagel Model became known to some as Footprint 1.0.

Overall the ecological footprint method focuses on measuring the environmental aspects which does not include economic or social aspects. Thus, if the companies focus on all three aspects of sustainability i.e., economic, social, and environmental should calculate their ecological footprint with their sustainability indicators.

Advantages

The ecological footprint calculation results in one number (in hectares or acres) that can be easily compared with other footprint calculations, whether at an individual, firm, local, regional, national or global level. Such comparisons can be useful for benchmarking across similar facilities or country operations (possibly normalizing the footprint per unit of production).

- The ecological footprint calculation can be used to complement other sustainability calculations, such as economic and social calculations. As such, the footprint calculation could form one section in a sustainability report.

Disadvantages

- Critics argue that aggregating the calculation of the ecological footprint into one number (in global hectares or global acres) oversimplifies diverse and complex environmental impacts;
- Because the ecological footprint focuses specifically on the number of hectares or acres required to support human resource demands, its calculations can place a higher value on some locations that might otherwise be valued less highly. For example, a productive mono-cultural agricultural crop can be valued higher than an ancient native forest or an organic crop that has a lower yield;
- The ecological footprint may identify densely populated cities as being more parasitic than rural areas because of their strong

reliance on the surrounding lands for their resources; and

- Despite the ecological footprint being applicable to all firms, some managers may not believe this method is relevant to their business because the calculations are based on farm, logging, and seafood businesses.

Calculating an Environmental Footprint

The Global Footprint Network published a document in October 2010. The document contained the data and formulas that can be used to calculate the environmental footprint (Ewing et al., 2010). Specifically, an organization can calculate the footprint of the environment and it can include the following:

- The area of cropland which is required to produce the crops;
- The land area for the purpose of grazing required to produce animal products;
- Forest area required to produce wood and paper products;
- The ocean area which is required to produce seafood;
- Land area is considered for housing and infrastructure; and
- Forest area required to absorb carbon dioxide emissions (Barrett and Scott, 2001; Collins et al., 2007; Dietz et al., 2007; Global Footprint Network, 2009; Knaus et al., 2006; Loh, 2003; Venetoulis and Talberth, 2009; Wackernagel, 1994).

Why Should Managers Care About This Method?

The ecological footprint method is popular since 1990 and hence there has been the introduction of newer tools such as water footprint, carbon footprint which are based on this methodology. This method is also older than any other method. This has been the focus of many studies. There is a great deal of information which currently exists to base the calculations. During the year, 2000–20003, the ecological tool was the most widely studied tool. However, it became the second most famous tool in the later years.

Many researchers favor this tool because of the results obtained. Since the results are in one number, measured in global hectares or acres, it indicates the environmental impacts. There are many other researchers who use this value to compare countries, regions or cities around the world. The ecological footprint of a multi-site organization in one location can be compared with its ecological footprint in other locations.

Therefore, the ecological footprint tool is mostly recommended for organizations that are operating in a variety of locations. The comparison must be made between the locations which have the similar line of operation.

2.2.6. Ecosystem Service Valuation

The services that contribute directly to life and include genetic and medicinal resources. These also include partial climate stabilization, natural disturbance regulation, raw materials, food production, erosion control, aesthetic beauty, human culture, recreation, preservation (including existence, bequest, and option values) and science and education, plant, and animal refugia, purification, and regulation of air and water, soil formation, detoxification, and decomposition of wastes, plant pollination, natural pest, and biological control, nursery function, seed dispersal, nutrient cycling, biodiversity maintenance, protection from the Sun's ultraviolet rays.

The area of concern is provided with a value on the ecosystem services. It helps in the determination of how the values and ecosystem services will change as a result of the implementation of a project or the enforcement of a land management policy. The ecosystem service valuation calculates the total economic value of the ecosystem goods and services in terms of the flows of benefits and costs provided by the stock of natural capital (Costanza et al., 1997; Daily, 1997; Daily et al., 1997; De Groot et al., 2002; Eftec, 2006; Kaval, 2010).

2.2.7. Environmental Evaluation Matrix

This method uses the adapted project life cycle framework. An environmental evaluation matrix is a tool used to assimilate environmental information to appraise projects. This tool also considers the "Design for Environment" principles (Graedel and Allenby, 1995; Labuschagne et al., 2005).

2.2.8. Environmental Input-Output Model

This is an economic tool that is used in describing the flow of goods and services in an organization that also considers the environmental effect. The tool helps in the identification of the interlinkage between the organizations as one organization uses the goods of others for the production or goods and services.

Figure 2.2: Wassily Leontief introduced the input-output method.

Source: Image by Wikimedia Commons.

The input-output model was introduced by Wassily Leontief in 1951 (Collins et al., 2007; Gay et al., 2005; Leontief, 1951; Wiedmann et al., 2007).

2.2.9. Environmental Management System Modeling

EMS is the tool that improves the environmental performance of an organization. After considering the short- and long-term environmental impacts of its goods and/or services, it focuses on providing the systematic methodology of managing an organization's environmental programs.

Under this process, there is the creation of an environmental policy statement; identifying significant environmental impacts; developing targets and objectives; implementing plans to meet those targets and objectives, including training employees to assist in fulfilling the environmental obligations; and reviewing and updating the process as necessary.

The typical EMS may include policies, task lists, data organization, audits, and environmental reports (Christini et al., 2004; Rao et al., 2006; Tam et al., 2006).

2.2.10. Epstein Roy Framework

This framework helps in the identification of ways in which the managers can affect sustainability performance and the corporate social and financial performance consequences. The Epstein Roy Framework is a comprehensive approach. The organization's manager can also create, apply or analyze their sustainability strategy (Epstein and Roy, 2001).

2.2.11. Genuine Wealth Accounting Model and Genuine Wealth Balance Sheet

The Genuine Wealth Accounting Model follows the traditional double-entry accounting conventions and encompasses five core capital assets. These core capital assets include social, manufactured, human, financial, and natural capital. Thus, the genuine wealth includes all those things that help in making the life worth living. Here both the assets are included, tangible as well as intangible like, strength of relationships with our friends and family, in addition to the ecosystem services that nature provides. Therefore, it can be said that it helps in entering and evaluating Genuine Wealth (Daly et al., 2005).

2.2.12. Green Globes Design

Green globes design is the tool or measure that helps in improving the sustainability and environmental performance of commercial buildings. The following elements are also included in this: an extensively detailed environmental assessment protocol, best practice guidance for green construction, software tools, qualified assessors that are experts in green construction, and a rating and certification system.

This method can be applied to new buildings, significant renovations, and the management and operation of existing buildings (Cole, 2006).

2.2.13. Green Productivity Index

The main focus of this measure is on the integration of environmental protection into corporate performance. It is calculated as the ratio of the selling price of the product to its production cost. While, as compared with its environmental impact, it includes solid waste generation, gaseous waste generation, and water consumption.

The purpose behind the measure is to improve the environmental decision-making and to promote environmental reporting and planning (Gandhi et al., 2006; Hur et al., 2004; Saxena et al., 2003).

2.2.14. Life Cycle Analysis

Under the life cycle analyses, measures the data in the quantitative term. It is also called a life cycle assessment. This measure evaluates the environmental and social damages related to a specific service or product in a quantitative term. Sometimes known as cradle-to-grave analysis or eco-balance.

It is measured while taking into consideration the entire life cycle of the service or product. This starts from product or service design to manufacturing, disposal, and all steps in between.

The International Organization for Standardization recommends the use of life cycle analysis procedures in corporate sustainability reporting (Hunter and Bansal, 2007; Karapetrovic and Jonker, 2003; Lopez et al., 2007b; Tam et al., 2006).

The life cycle analysis is also called the life cycle assessment. The simplified life cycle analysis was found to be the most commonly studied tool in the study's database. For the first time, Coca-Cola company used the life cycle analysis tool in the year 1969. They conducted the study to know that which variety of container of Coca-Cola had the smallest effect on the environment.

After the company tested the method, other companies also started to use this tool in the United States and Europe on their products and services (Scientific Applications International Corporation, 2006). The study revealed that there is an increase in the popularity of the studies investigating the life cycle analysis method over the past 10 years.

The organizations can make their own choices by using the life cycle approach. This method lets the organization make informed choices after considering all the links of the life cycle chain. This also includes those links that are outside the purview of the company.

A life cycle analysis follows a product or service throughout its entire life cycle. It starts from raw material acquisition to manufacturing, production, use, reuse, maintenance, and waste management.

There are four basic steps for conducting a life cycle analysis:

- Scope and goal definition;
- Inventory analysis;
- Impact assessment; and
- Interpretation of results.

Figure 2.3: Steps in conducting life cycle analysis.

A comparison of all possible social and environmental damages related to the product or service under review are possible in the life cycle analysis. There are some danger categories that is evolved during the study. The danger categories include, greenhouse gases, soil acidification, water pollution, ozone (O_3) depletion, and habitat destruction.

The information obtained after using the life cycle analysis can be used to select the least ecologically, or least socially, burdensome of the analyzed products or services. The life cycle analysis is as good as its data. Under this method, the use of the current method is essential. For example, for the results to remain relevant, the products are needed to be updated every month in the case of the phone and computer products. The change is so frequent that a life cycle analysis for these is not possible.

The United States Environmental Protection Agency (EPA) website shows that there are more than 30 organizations that have created a software package for conducting the life cycle analysis.

The procedures of life cycle analysis are also part of the International Organization for Standardization ISO. According to some, a simplified life cycle analysis is a more realistic type of life cycle analysis because it addresses aspects directly related to a product.

In order to make a linkage between the society and the environment, there is a related tool that helps. The tool is known as materials flow analysis which is also known as substance flow analysis. This method analyzes its effect on the ecosystem and the flow of materials through an industry or firm. Therefore, materials flow analysis is directly comparable to a life cycle analysis (Hobbes et al., 2007).

Advantages

- The conduction of new life cycle analysis is easier when a firm has already constructed a model of life cycle analysis. It is better than comparing the analysis for a variety of products or services;
- This analysis is helpful in making choices for a firm and in analyzing important external factors (e.g., political, economic, technological, and social);
- There are many software packages that are available to assist a company in conducting a life cycle analysis. There are some software packages that are available at very little or no cost to the user;
- There are many life-cycle analyses that do not include economic values. However, there is a tool that is available which combines input-output analysis with the life cycle analysis called EIO-LCA (Environmental Input-Output Life Cycle Analysis); and
- Because the tool is much popular, extensive information is available to support the firms in conducting a life cycle analysis. The tool supports the firm from inventory data to assessment information and assessment tools. The organizations have set up in a way as to provide the entire life cycle analysis to assist other organizations.

Disadvantages

- There is a common saying in this, "the analysis is as good as the data." Under the life cycle analysis tool, the company may face difficulty while observing the product from its beginning to its disposal. The difficulty is in obtaining the data of a product, when they obtained a required material and also after it leaves the hands and the consumer either disposes of it or recycles it; and
- There are different ways in which the life cycle analysis is done. There is no single method that the tool uses. The results cannot be compared across two or more life cycle analysis studies due to the lack of uniformity.

Why Should Managers Care About This Method?

This method was first used in the late 1960s and the popularity of this method is increased between the year 2000 and 2010. Since the method is very popular and it has been included in the International Organization for Standardization environmental standards organizations will find that useful information is available on the lifecycle analysis.

The life cycle analysis is well assessed in Europe, North America, and Oceania. Along with the online journals and guidelines, firms can access more than 30 computer applications, many of which are free. The life cycle analysis can be conducted in many ways, however, this tool studies mostly the new product or when assessing a specific event.

2.2.15. Lowell Center Hierarchy

This tool helps in the measurement of the effectiveness of an organization's sustainability principles by measuring their progress. The hierarchy was developed at the Lowell Center for Sustainable Production, University of Massachusetts Lowell (United States). The assumptions are based on five-level hierarchy that includes compliance/conformance, resource use and performance, local effect indicators, upstream, and downstream/supply-chain, and life cycle indicators and sustainable systems indicators (Gerbens-Leenes et al., 2003; Veleva et al., 2003).

2.2.16. Materials Flow Analysis

Materials flow analysis is also called as substance flow analysis. This measure is similar to a life cycle analysis and is used in the area of industrial ecology. In order to study the linkages between society and the environment, this tool is useful. This tool helps in the analysis of both the flow of a material through an industry or firm and the affected ecosystems. It gives results on a national scale, a regional scale, a corporate or industrial scale or during the life cycle of a product (Hobbes et al., 2007).

2.2.17. Responsive Business Scorecard

In view of the improvement in the performance of the companies. This responsive business scorecard is a useful tool used by the companies to integrate the stakeholder's demand. The companies try to improve their performance based on environment, social, and financial goals.

As per the tool, the most common perspectives it takes into considerations are suppliers and customers, the plant and society, internal processes, owners, and financiers, and learning and employees (Caldelli and Parmigiani, 2004; van der Woerd and van den Brink, 2004).

2.2.18. Sustainable Value Added

Under the sustainable value-added tool, simultaneously three aspects are considered i.e., economic, environmental, and social. All the forms of capital are kept constant and are inspired by the concept of strong sustainability. This tool measures the extra value created by a company, when the company becomes more efficient. It also makes sure that all the environmental and social impacts remain at a constant level.

This tool considers eco-efficiency and social efficiency at the same time (Figge and Hahn, 2004).

2.2.19. Triple Bottom Line Reporting

This tool is the dominant approach which was recommended by the United Nations standard in the year 2007 for urban community accounting created. This tool uses public sector full-cost accounting. This method expands on a typical economic accounting method and includes ecological and social aspects.

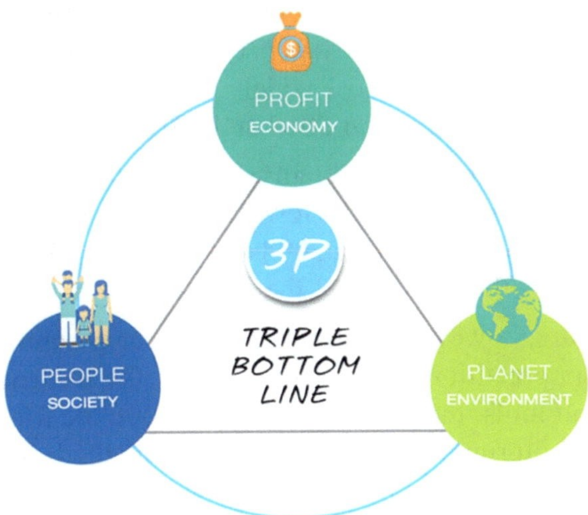

Figure 2.4:3Ps in triple bottom line (TBL).

Source: Image by Sustainability.

As per the concept of TBL reporting, the stakeholders are ones on which the company's responsibility falls. This should not be confused with the shareholders. Therefore, the interests of the stakeholders are more important than maximization of the profits of shareholders (Ballou et al., 2006; Elkington et al., 2006; New Belgium Brewing Company, 2009; Schafer, 2005).

2.2.20. Whole Life Value

The whole life value tool works with the life cycle analysis. It is a tool that considers the life cycle analysis with that of the value of the clients and other stakeholders. These client values include the aspects of the project delivered and the definition of the sustained use aspects. This tool helps in the determination of how the stakeholders can be considered while the evaluation of the facility process (Holt, 2001). Moreover, it helps in identifying those stakeholders.

2.3. TRENDS OVER TIME

Since the year 2004, the most widely used and important tool for the measurement of the environment impacts is the life cycle analysis. However, the ecological footprint was used mostly in the year 2000–2003. The ecological footprint was most widely recognized as the second most commonly studied tool in the year 2008 and 2010.

The carbon footprint and ecosystem service valuation tools seem to have increased their popularity in the last few years. These tools were the second and third most popular tools during 2008 and 2010, respectively. It therefore implies that these methodologies are the emerging methodologies.

These tools are, however, separated by the year grouping and the study type (academic or practitioner). According to this, the academicians, favored two measurement tools: the ecological footprint and the life cycle analysis. The ecological footprint was the most popular method during the period of 2000 and 2003. Whereas, the life cycle analysis was the most popular method between 2004 and 2010 (Figure 2.5).

While, the practitioner literature, the most popular tools were the ecosystem service valuation and the life cycle analysis, one valuation-based tool and one measurement-based tool, respectively for the year 2008 and 2010.

2.4. POPULARITY OF METHODOLOGIES BY ANALYSIS LEVEL

The study of the analysis level also revealed that which tools were the mostly used at each level of analysis. Mostly, a large proportion of study examined the entire organization, specific product or event. Therefore, more information was available for these categories.

The entire organization's environmental performance was measured and found that there were four tools that stood out: the balanced scorecard approach, life cycle analysis, EMS modeling, and the ecological footprint.

While in the case of the environmental performance of a specific product or an event, the most common studied tool was the life cycle analysis. It studied the environmental impacts of a city, region or a country.

In the study of ecosystems, the ecosystem service valuation was found to be most popular tool.

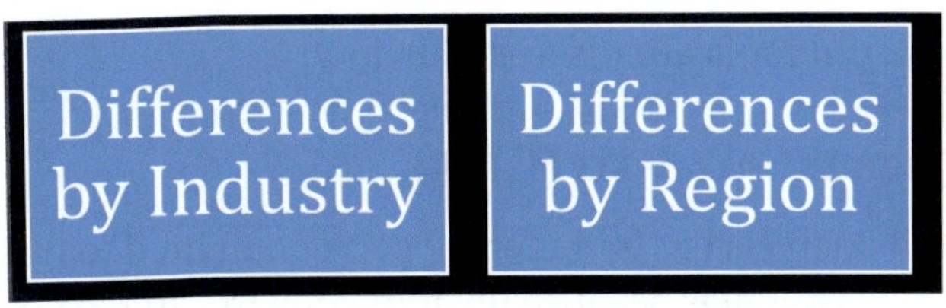

Figure 2.5: Differences in tools based on industry and region.

2.4.1. Differences by Industry

The tools are differentiated based on the industry type. In this, industries are grouped into nine categories: basic materials, consumer goods, financial, health care, industrial goods, services, technology, utilities, and there is an unspecified category that includes all studies that did not identify a specific industry. The tools after separation were not appeared to be favored.

However, there were two exceptions in this, the consumer goods and the industrial goods sectors, which clearly favored the life cycle analysis tool. Although, other tools were studied in those industries. Other tools did not appear in more than three studies within any specific industry category. There is a column labeled "unspecified" which was created for the studies that did not identify a specific industrial sector. This category revealed the balanced scorecard, ecological footprint, ecosystem service valuation, and life cycle analysis were the tools that appeared in three or more studies.

2.4.2. Differences by Region

As per the differentiation based on regions, Asia, Europe, North America, and Oceania were the four main regions where the life cycle analysis was a commonly studied tool. The database published two or more studies of this tool. In Asia and North America, the EMS model was investigated in the studies. while the ecological footprint model was investigated in both Europe and North America.

Whereas the life cycle analysis, ecological footprint and environmental input-output model were applied in three or more studies in Europe. In the other three continents, these tools were investigated in two or fewer studies. However, assumptions could not be made based on the small number of studies published in Asia, North America and Oceania.

2.5. VALUATION-BASED TOOLS

The valuation-based tools are discussed below:

2.5.1. Cost-Benefit Analysis (CBA)

The cost-benefit analysis has a bigger scope as compared to other valuation-based tools. This is popular because the tool converts all the cost and benefit into monetary terms. Hence, this tool not only compares the programs within the health care but also can assist in the decision making both within and outside sector of the economy (Drummond, Sculpher, Torrance, O'Brien, and Stoddart, 2005).

Figure 2.6: Cost-benefit analysis.

Source: Image by Picpedia

The cost-benefit analysis (CBA) informs questions of allocative efficiency as it assigns relative values to goals in health and non-health sectors. This helps in determining which goal is worth achieving, and which program are worthwhile doing given the alternative use of resources (Drummond et al., 2005).

2.5.2. Ecosystem Service Valuation

Ecosystem service valuation is a tool that contributes directly to the life. It includes the dispersal of seeds for the flora reproduction, a conservation area that supplies both recreation to humans and a habitat for animals. There are total of 22 categories that the ecosystem comprises (Kaval, 2010)

Ecosystem service valuation is basically the process of placing values on the ecosystem services in the area of concern. This also includes the determination of how the values and ecosystem services will change as a consequence of a project. These include the construction of a new building or the clearing of land to plant agricultural crops.

The process is complicated as placing the value is a difficult task. This may include the simple calculation of the price of gold to the more complicated calculations which can include the value of bees pollinating crops (Merlo and Croitoru, 2005; Pearce and Turner, 1990).

The calculation of all the ecosystem services yields a total economic value or a flow of all benefits and costs provided by the ecosystem. The total economic value found is the complete analysis as it considers the values that has already been used or the value of actually using a resource in a particular way (e.g., fishing). This is in addition to all the non-use or indirect values when the resource is not used directly which is still a benefit e.g., the value of worms churning the soil of crops) (Costanza et al., 1997; Daily, 1997; Daily et al., 1997; De Groot et al., 2002; Eftec, 2006; Kaval, 2010).

The ecosystem valuation method can be categorized into two major groups: market values and non-market values. The out of pocket expenses which are traded in the formal market such as the sale of fish caught or the cost of monthly electric bill. Whereas the non-market values are those which are not easy to calculate.

However, there are various tools that are available to calculate this value that includes the travel cost method, the hedonic pricing method, the contingent valuation method, the choice modeling method, the avoided cost method, the restoration cost method, the replacement cost method, the factor income method and the benefit transfer method.

The ecosystem service type is different and there are different valuation methods that are applied based on the relevant to the project being studied. For any given service, there are many different that can be applied, and the choice of method depends on the project goals and availability of resources, funding, and time.

Advantages

- The ecosystem service valuation provides the most in-depth perspective possible, after a complete ecosystem service valuation is conducted. It is thoroughly conducted through valuation;
- The extensive information for an organization is received when it is considering the construction of a building, restoring an area (such as returning as a former mine to its original condition) or another land use change; and
- This method makes it easy for the extractive industries to visualize and calculate their impacts.

Disadvantages

- The ecosystem valuation method conducts a complete ecosystem service valuation which is an extensive process. This process is time consuming and many resources are required. Also, the process can be expensive. A firm may thus choose to focus only on the ecosystem services of greatest interest; and
- Comparison between the studies are not always possible because each ecosystem service is valued in many ways and researchers use different techniques to value the different ecosystem services.

Calculating an Ecosystem Service Valuation Study

While calculating an ecosystem service valuation method, there are two studies that are helpful for organizations that are interested in conducting an ecosystem service valuation study: Kaval (2010) and Riera et al. (2011). One is the Kaval study that provides step-by-step instructions for conducting an ecosystem service valuation. Other being the Riera et al. (2011) publication focuses on extensive details of the various valuation methods, such as calculating a contingent valuation study.

Why Should Managers Care About This Method?

Ecosystem service valuation is the valuation-based method that is gaining popularity in these days. When an organization has plans of modification in terms of piece of land in some way or the other, this method is the most used method. This may include the conversion of a forested area into a mine or by constructing a new building. This is because if the building is constructed or the mine has been evacuated, it may not be possible to determine the original ecosystem prior to the land change.

This tool is also helpful in restoring the area to a condition closer to its original condition in an organization. This is due to the reason that the ecosystem service valuation is a new tool and the managers, and the researchers needs to determine which aspects and values to focus on. After a complete ecosystem service valuation has been conducted, applying the same procedure to new sites will be easier because the tools will already be in place.

Managers may find it useful for building an evidence-based defense in the case of environmental litigation. This also includes studying the impact of new environmental regulations that may require changes in company operating procedures. Because ecosystem service valuation requires extensive documentation of all environmental effects.

2.5.3. Environmental Input-Output Model

An environmental input-output model is a valuation-based tool. This tool is helpful in describing an organization's tool flow of goods and services, including their environmental effects.

This tool also helps in determining that how organizations are linked together. Because organizations use the products and services from other organizations to produce their own goods and services. The input-output model was introduced by Wassily Leontief in 1951 (Leontief, 1951).

The relationship between different departments, divisions or companies within the organization is demonstrated in the environmental input-output model. This is in the addition to the relationship the organization has with other organizations.

In simple terms, the input-output model is based on the following equation:

Production – Consumption = Demand.

In relation to the production and consumption, numbers that correspond to the inputs and outputs are entered into a matrix. This illustrates how outputs change when inputs change or, more specifically, how profits change when demand or the costs of inputs (including the costs of environmental pollution) change.

There are two basic players in an input-output model: the producers and the consumers and/or users. The producers are the creators of the products. For example, an administrative department both hires employees for the manufacturing department and they also market the product the manufacturing department produces; the raw material supplier sells raw materials to the manufacturing department for the use in their product development.

The office supplier provides the administrative department with the tools it requires, such as paper, copy machines, pens, and computers. The consumers and users are not only the people who purchase the products and use the final products but also other users, such as the staff in the administrative department who use the computers they obtained from the office supplier and the employees of the manufacturing department who use the raw materials they obtained from the raw material supplier to produce the final goods.

2.5.3.1. Calculating an Environmental Input-Output Model

Conway-Schempf (1999) has conducted a detailed report that elaborates how an organization can conduct an input-output study (Conway-Schempf, 1999). The United Nations System of National Accounts also provides a detailed explanation on calculating input-output analysis in its Handbook of National Accounting: Integrated Environmental and Economic Accounting 2003.

2.5.3.2. Why Should Managers Care About This Method?

An organization which wants to create an environmental input-output model would essentially create a spreadsheet of all the project's benefits and costs, in addition to the outlining the effects to the entire system if one of the benefits or costs were to change.

For example, for producing 10,000 more widgets, the spreadsheet should include the benefits of selling more and the administrative, manufacturing, and environmental costs of producing more. The environmental input-output model can be applied to any output that can include a product or a service. There are many resources available to instruct the new user in calculating an input-output analysis, simplifying the process for firms attempting to apply it.

However, there is an extremely detailed input-output analysis which is necessary, the experts may be needed to assist in matrix creation. Once the matrix is created, however, the process is again simplified for the manager.

Advantages

- This tool helps in illustrating the effect on profits, outputs, and environmental pollution as various inputs change. This tool helps in determining all the organization's activity;
- The input-output tool is easy to use and also straightforward due to the basic calculations involved. However, for an extremely complicated input-output table, an expert may be needed to create the matrix. Although, once the matrix is created, the process is again simple for the manager as they will just enter numbers into the matrix and the rest will be automatically calculated;
- There is a System of National Accounts created by the United Nations to help countries to collect data for input-output analysis and as a result, enable easier international comparisons;
- There is an extensive analytical power in the use of input-output model; and
- Input-output models present an extensive amount of data in an easy-to-understand and orderly fashion.

Disadvantages

- There is a requirement of an extensive data in this method because all costs and benefits for every economic activity must be represented;
- There may be variety in the quality of the data because many organizations may not maintain records of all the data required;
- An organization may face difficulties obtaining data from all departments, divisions or companies;
- There is a difficulty in obtaining data from other organizations they are involved with, especially in relation to environmental pollution;
- Experts are needed to create the matrices because of the extensive data requirements;

- The data is needed to be updated constantly due to the accuracy, because technology, demand, and prices are constantly changing; and
- Input-output models believe the general assumption that a firm has one production function and produces one product, which is typically not the case.

2.5.4. Sustainable Value Added

Sustainable Value is a valuation tool that helps in assessing the sustainability performance of companies using the opportunity cost thinking. The financial, environmental, and social resources of the company are assessed and seen that what creates more return for the sustainable value of a company (or other organizational entities). As a result, with the set of economic, environmental, and social resources compared to a benchmark, Sustainable Value shows how much more or less return a company created.

2.5.5. Triple Bottom Line Reporting

TBL Reporting focuses on the point that instead of one bottom line there must be three bottom lines, i.e., profit, people, and planet. John Elkington coined the term TBL. Elkington first measured the sustainability during the mid-1990s by encompassing a new framework to measure performance in corporate America.

The accounting framework called the TBL, is beyond the traditional measures of profits, return on investment, and shareholder value to include environmental and social dimensions. The TBL reporting can be an important tool to support sustainability goals by focusing on comprehensive investment results. This is with respect to the performance along the interrelated dimensions of profits, people, and the planet.

People nowadays are taking more interest in TBL accounting for-profit, nonprofit, and government sectors. There are many businesses and nonprofit organizations that have adopted the TBL sustainability framework. They evaluate their performance, and a similar approach has gained currency with governments at the federal, state, and local levels.

2.6. OTHER MEASURES

Environmental impact methodologies can be categorized into the interaction matrices and checklists. The methodologies are the networks and energy system diagrams representing the variations of interaction matrices. Interaction matrices calculate simple considerations of project activities and their impacts on environmental factors. This also calculates the stepped approaches which display interrelationships between impacted factors.

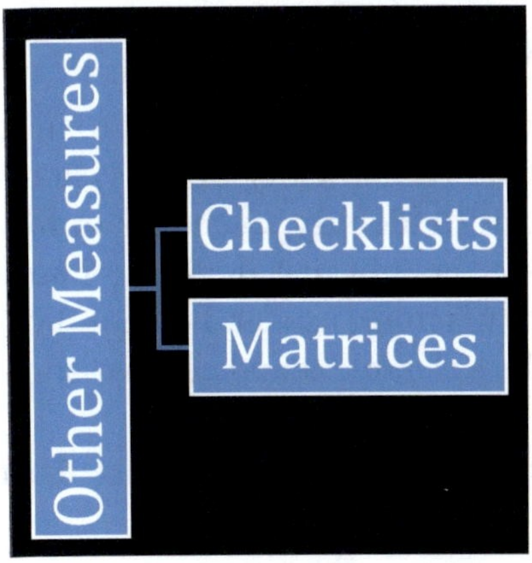

Figure 2.7: Other measures that are used for the measurement.

Checklists include simple listings of environmental factors and the descriptive approaches which include information on measurement, prediction, and interpretation of changes for identified environmental factors. Checklists also include the scaling or ranking of impacts of alternatives on each of the environmental factors under consideration.

Scaling techniques are useful in the use of numerical scores, letter assignments, or linear proportioning. There are alternatives that can be ranked from best to worst in terms of potential impacts on each environmental factor. Those involving the assignment of importance weights to environmental factors and the scaling or ranking of the impacts for each alternative on each factor are the most sophisticated checklists.

The development of a product mix enables the comparisons which consist of multiplying importance weights by the scale or rank for each alternative. The methods can be useful throughout the impact assessment process, with certain ones being of greater value for specific activities.

More particularly, matrices and networks are useful for impact identification. On the other hand, weighting-scaling checklists find the greatest application in the final evaluation of alternatives and the selection of a proposed action.

2.7. THE LIMITS TO MEASUREMENT

There are two essential characteristics associated with the measurement and assessment of the environmental impacts by the products and materials:

2.7.1. The Data is Incomplete

The ecosystem is so complex that it is difficult to make the cause and effect relationship. The scale and complexity of ecosystems make long chains of cause and effect difficult to identify, even less to measure. However, there is a common acceptable thought that the more immediate the calculations associated with the manufacture, use, and disposal of the product, the more is the sound base for the assessment and comparison.

2.7.2. Some Values Are Considered More Important Than Others

Since there is a lack of environment-based data. There is a complexity of the impacts which are hard to model and difficult to understand. There is much of the grey areas that is a subject of debate. When there is a need to draw a risk matrix, which is crude form of risk management, uncertainty is evident.

The risk in measured in terms of scale. The scale of risk is the feature of ecolabel schemes which seek to estimate the importance of the environmental impact in relation to the others in the assessment. While the importance of risk is drawn, 'weighting' is used to modify the data. There is a common belief of people that says that global warming is chief amongst the risks, but the relative importance of other impacts is contentious.

There can be conflicts with the other more familiar non-environmental values due to the pragmatism of construction. There is an increasingly common encounter which is the design of insulation for the retrofit of solid-walled homes. The most acceptable solution is not always the best environmental outcome: Example: Retrofitting internal v external insulation.

The most efficient way is to apply insulation to the outside face of an external wall. In this instance, environmentally conscious designers might look to resource efficiency as a very important value with no planning constraints. Moreover, the specification can be there of wood fiber insulation from renewable sources for external application.

However, if there are constraints related to the planning process, such as a listed building or location in a conservation area, external retrofitting is problematic. This situation calls for the only possible intervention that is to apply insulation to the inner face of the wall. Those familiar with refurbishment are aware that floor areas are usually at a premium and not all people welcome the thickness of wood fiber insulation eating into their rooms.

In order to provide the identical overall u-value, the specifier will look into a more 'dimensionally-efficient' product which is usually a non-renewable oil-based material such as polyurethane or phenolic foam. This way, the value of preserving space for occupants has effectively 'trumped' the value of resource efficiency and the pollution associated with petrochemicals.

2.8. CONCLUSION

There must be two-goal in any efforts to limit human environmental impacts. One being the minimization of the undesirable environmental impact per unit of human activity and the other being to ensure that the cumulative impact of all human activities is compatible with the persistence of all critical ecosystem conditions and processes.

However, these goals are difficult to achieve as there must be profound uncertainties that will complicate efforts to achieve these goals, but they are the appropriate targets. The measures of the environmental impacts are the key tools in identifying the opportunities in moving towards the first of its goal.

The measures of ecosystem condition are vital in charting its progress toward the second goal. These two sets of tools are the most useful. If they can be used to accurately predict the consequences of human activity on affected ecosystems.

In order to achieve this, they must be refined and coordinated. This should be done in such a way that predictions can be based on the product of the environmental impact per unit of an activity multiplied by the scale of that activity. This information will be useful for distinguishing acceptable and unacceptable impacts.

As per the environment impact assessment, there are two broad categories to study the environmental impact. These are the measurement tools and the valuation tools. These are discussed in detail in this chapter. Concluding this, there are total 15 measurement tools and 5 valuation-based tools.

REFERENCES

1. Kaval, D., (n.d.). Measuring and Valuing Environmental Impacts. [eBook] Available at: https://static1.squarespace.com/static/5d5156083138fd000193c11a/t/5d6414993db3970001f7c764/1566839962975/NBS-Systematic-Review-Impacts1.pdf (accessed on 4 March 2020).
2. Nap, (n.d.). Measures of Environmental Performance and Ecosystem Condition. [online] Available at: https://www.nap.edu/read/5147/chapter/2#11 (accessed on 4 March 2020).
3. Slaper, T., & Hall, T., (n.d.). The Triple Bottom Line: What Is It and How Does It Work? [online] Ibrc.indiana.edu. Available at: https://www.ibrc.indiana.edu/ibr/2011/spring/article2.html (accessed on 4 March 2020).
4. Sustainablevaluecalculator.com. (n.d.). Sustainable Value-How to calculate Sustainable Value. [online] Available at: http://www.sustainablevaluecalculator.com/en/info/calculation/ (accessed on 4 March 2020).
5. Syafinaz S., I., Hanafiah, J. M., Ibrahim, F., & Abdul, M. R., (2017). *Valuation of Impacts in Cost-Benefit Analysis.* [online] Available at: https://www.researchgate.net/publication/319953004_VALUATION_OF_IMPACTS_IN_COST_BENEFIT_ANALYSIS/link/59c332230f7e9b21a82ab021/download (accessed on 4 March 2020).

CHAPTER 3

Environmental Solid Waste Auditing and Management

CONTENTS

In the chapter, environmental solid waste auditing and management, various key terms related to solid waste are discussed in order to get better insight about how auditing and management of solid waste is possible. The chapter discusses the meaning of solid waste and various types of solid waste.

The chapter also discusses the types of municipal solid waste and how to manage the municipal solid waste. The chapter also highlights environmental and health risk scenarios related to solid waste. The chapter also discusses the legal framework applicable to municipal solid waste management (MSWM). In addition, it also discusses various methods of solid waste management (SWM).

The chapter also describes the functional elements of the waste management system. The chapter also explains one important concept which is integrated solid waste management (ISWM). The chapter also recognizes the importance of understanding the health effects of incineration. The chapter also discusses some of the hazardous substances associated with waste management. In the end, the chapter explains the role of sustainable SWM.

3.1. INTRODUCTION

Waste is a continually growing problem at local, national as well as international level. Solid wastes are mainly the result of human and animal activities that are normally discarded as useless or unwanted. In other words, solid wastes can be defined as the organic and inorganic waste materials that are generated through various activities and actions undertaken by the society and the waste that have lost their value to the first user.

After significant development in the industrial production and consumption activities, urban society rejects and generates solid material at a more rapid rate which results in a significant increase in the volume of waste generated from several sources such as, domestic wastes, institutional wastes, commercial wastes, and industrial wastes of most diverse categories.

Management of solid waste is concerned with having control on generation, storage, transfer, collection, and transport, processing, and disposal of solid wastes in a systematic manner that is in accord with the best principles of public health, engineering, economics, aesthetics, conservation, and other environmental considerations.

In its scope, SWM is the combination of various activities that include administrative, legal, financial, planning, and engineering functions aimed at finding solutions to problems of solid wastes thrust upon the community by its inhabitants (Tchobanaglous et al., 1997). Solid wastes as the name suggests has the potential to pollute all the vital components of the living environment (i.e., land, air, and water) at both national and international levels.

The problem is compounded by the current trends in production and consumption patterns and by continuing urbanization of the world. These problems are more prevalent in developing countries compared to developed ones as the economic growth as well as urbanization is more rapid.

3.1.1. Solid Waste Management

Management of solid waste is concerned with regulating and controlling of generation, collection, storage, transport, and transfer, processing, and disposal of solid wastes in such a way that it is in accordance with the best principles of economics, public health, conservation, engineering, aesthetics, and other environmental considerations. In its scope, it includes all financial, administrative, planning, legal, and engineering functions involved in the whole spectrum of finding solutions to variety of problems of solid wastes thrust upon the community by its inhabitants (Tchobanaglous et al., 1997).

3.1.2. Categories of Solid Waste Solid

Table 3.1: Waste Can Be Categorized Based On Source

Source	**Typical facilities, activities, or locations where wastes are generated**	**Types of Solid waste**
Agricultural	Field and row crops, orchards, vineyards, diaries, feedlots, farms, etc.	Spoiled food wastes, agricultural wastes, rubbish, and hazardous wastes
Industrial	Construction, fabrication, light, and heavy manufacturing, refineries, chemical plants, power plants, demolition, etc.	Industrial process wastes, scrap materials, etc.; nonindustrial waste including food waste, rubbish, ashes, demolition, and construction wastes, special wastes, and hazardous waste.
Commercial and Institutional	Stores, restaurants, markets, office buildings, hotels, auto repair shops,	Paper, cardboard, plastics, wood, food wastes, glass, metal wastes, ashes, special wastes, etc.
Municipal solid waste	Includes residential, commercial, and institutions	Special waste, rubbish, general waste, paper, plastics, metals, food waste, etc.

Source: Table from wgbis.ces.iisc.ernet.in

3.1.3. Municipal Solid Waste

The term municipal solid waste (MSW) is generally believed to comprise all the waste the that is generated by the community over a period of time, with the exception of waste generated by treatment plants, municipal services, and agricultural and industrial and processes (Tchnobanoglous, G, and Kreith, F., 2002). In the context of urban area, the term municipal solid waste is of greater importance. The term refers to the total amount of waste collected and regulated by the municipality and comprises of most diverse 13 categories of wastes. It basically includes wastes from various categories such as, domestic wastes, institutional wastes, commercial wastes, and building materials wastes.

3.1.4. Types of Municipal Solid Waste

Table 3.2: The Sources of Municipal Solid Waste

Sources	Examples
Residential	Single family homes, duplexes, town houses, apartments
Commercial	Office buildings, shopping malls, warehouses, hotels, airports, restaurants
Institutional	Schools, medical facilities, prisons
Industrial	Packaging of components, office wastes, lunchroom, and restroom wastes (but not industrial process wastes)

Source: Table from wgbis.ces. iisc.ernet.in

3.1.5. Municipal Solid Waste Management

MSWM includes the application of the principle of Integrated Solid Waste Management (ISWM) to municipal waste. ISWN is the application of suitable technologies, techniques, as well as management programs that encompasses all types of solid wastes from all sources to achieve the twin objectives of (a) minimization of waste and (b) effective and efficient management of waste still produced after waste reduction.

In the MSWM, some of the issues that should be given higher importance are:

- Increasing waste quantities;
- Wastes which are not recorded in the national municipal solid totals;

- Lack of clear definition about SWM terms and functions;
- Limited data availability as well as data quality concerns;
- Need for transparency in operations and policies and strategies adopted and implanted by the state and local government; and
- Need for even and predictable enforcement regulations and standards.

3.1.6. Environmental and Health Risk Scenarios

There are countless hazards of solid wastes to the living community if it is not managed properly. Solid wastes possess the ability to pollute all the essential elements of living environment (for instance air, water, and land). Some of the hazards caused by solid wastes are discussed as follows:

- The wastes which are often uncollected end up in drains, causing blockages which may contribute to flooding at the time of heavy rains and also become the reason of unsanitary conditions.
- Open and overflowing bins attract stray dogs which is the root cause of spreading animal related diseases such as the spread of rabies.
- Open waste bins also attract stray and domestic cattle. This kind of scenario is not common in the developed countries, however in developing countries and less developed countries such as India, it is clearly observed in every part of the country. Cattle in the city is responsible for causing nuisance by blocking the traffic on the roads. Cattle that graze on the waste from bins ended up consuming along with the vegetable matter, which may result on life threatening for them. The milk which is delivered by cattle that feed on waste can be contaminated and may also prove unsafe for human health.
- Flies breed in few ingredients of solid wastes, and as flies roam all around, they spread the disease more rapidly born from the solid waste.
- Mosquitoes breed in rainwater and blocked drains that is retained in discarded tire, cans, and other objects. Mosquitoes spread disease, including dengue and malaria.
- Rats find shelter and food in waste dumps. They play the role of spreading the disease as they eat the spoil food and spread the disease to different areas.

- In developing countries, it is also seen that because of lack of wastage handling techniques, people generally burn the waste, which is the easy way to get freedom from the waste. However, it is the significant contributor of air pollution; the products of combustion include dioxins that are particularly hazardous.
- Dusts and Aerosols can spread pathogens and fungi from decomposing and uncollected wastes.
- Uncollected waste degrades the urban environment conditions, make it a challenging task to keep streets and open spaces in a clean and hygienic condition. Plastic bags are in specific the largest contributor of creating an aesthetic nuisance.
- Workers of waste collection workers face various occupational challenges, including strains from lifting, contact with pathogens, and injuries from sharp objects when manually handling the waste.
- Dangerous items (such as razor blades, broken glass, hypodermic needles and other healthcare wastes, potentially explosive containers, and aerosol cans and chemicals from industries) may pose potential risks of injury or poisoning, majorly to children and people are more vulnerable to this kind of waste.
- Heavy refuse collection trucks may pose a serious threat to the surfaces of roads that were not intended at holding such huge weights.
- Waste items which are reused without being completely cleaned or sterilized can spread infection to later users. (Examples are bottles and medical supplies.)
- Polluted water (leachate) flowing from disposal sites and waste dumps can cause hefty pollution in water supplies, lakes as well as ponds. Chemical wastes (particularly persistent organics) may prove to be fatal or have serious effects if inhaled, ingested, or touched and can cause extensive pollution of water supplies.
- Liquids and fumes, release from deposits of wastes (perhaps formed as mixture of hazardous chemical between components in the wastes), can prove to be fatal or have serious effects.
- Methane (one of the key components of landfill gas) is much more hazardous comparative to carbon dioxide in context with contributor to greenhouse gas, leading to climate change.

- Fires on disposal sites may cause significant air pollution, causing illness, spreading disease and at the same time reducing visibility, making disposal sites riskily unstable, causing bursts of cans, and possibly spreading to adjacent property.
- Unlike developed countries where there is proper management system, in developing countries, there are rag-pickers working on landfill sites that are prone to many diseases, for instance, respiratory infections such as lung impairment.

In order to know the impact of SWM, especially on developing countries, a study carried out by Chittaranjan national Cancer Institute, Kolkata to study the health of Delhi's rag pickers with that of the control subjects from east Delhi slums.

Nearly 75% of rag pickers from the sample group had higher frequency of upper respiratory symptoms (such as sinusitis, sore, running, or stuffy nose, common cold, throat, fever) and 81.6% showed lower respiratory symptoms (such as dry cough, wheezing, cough with phlegm, and chest discomfort) and breathing problem.

However, in the developed countries, there are numerous technologies/ options are available, which prevent from these kinds of diseases. Replicating the same in low-income countries is incompatible/ inappropriate. The success of waste disposal practices depends largely on overcoming the following constraints,

- **Municipal Capacity:** The scale at which this operation operate of task is giant and the concerned municipal department are able to collect just 60 to 70% of total waste generated, so treatment and disposal activity inevitably receives less attention. There are various efforts initiated by the municipal department in order to address this lack of capacity for instance, by allocating this operation to the private sector.
- **Political Commitments:** Solid waste management is far more than just a technical issue; it is a complicated task that has various implications for employment, local taxation, and regulation of public and managing authorities. Any change in this area required the need of political interference in order to implement the effective policies and strategies.

No major action can be taken without the involvement of politics. As SWM is for the interest of country, political commitment become obvious. However, it is rarely a topmost priority for political organizations unless it carries and attract strong and active public interest. This is viewed as a cost to the "public" without apparent returns.

- **Finance, Cost Recovery, and Resource Constraints**: Ensuring the proper and effective management system represents a huge and consistent investment and it may be therefore challenging to give it priority over other resource demands.

Majority of the waste management department are short of enough funds and resources in order to carry out their operations and activities smoothly. Since the collection and transport, itself generally direct SWM costs in developing countries, safe disposal perpetually receives less emphasis whereas in all other developed countries concentrate on all aspects of management.

- **Technical Guidelines:** Unlike the higher income countries groups, low-income countries cannot match their standards of planning, and strategies implementation primarily because of the difference in climate, institutions, resources, foremost priorities, etc. However, in low-income countries, some sort of little adequate guidance is available. Arising from this uncertainty, officials find themselves ill equipped to plan management strategies, that can be achieved easily and can play a crucial role in avoiding unacceptable environmental hazards.
- **Institutional Responsibilities:** Though effective management of the wastes is the duty and responsibility of the municipality, there is no clearly stated vision of management (for instance, enough recognition is not given to SWM). Existing vision followed by a typical apathy to solid waste is an "out of sight is out of mind" attitude by the municipalities, concerned organizations and public because of lack of implementation of tough rules and regulations just like as in the prevention of emission of water and air pollutants.

Proper management of solid waste requires the need to ensure communication, cooperation, and communication among all the concerned authorities and also include organizations that are accustomed to acting independently but the lack of accountability in all levels of management.

Among the authorities, there is a need to ensure clarity and transparency in the roles and responsibilities assigned to the people of different departments. Some smaller towns may not have human resources with particular skills for providing a SWM service.

- **Inadequate Legal Provisions:** It is generally seen that in the majority of the countries, the laws and regulations on SWM are fragmented and outmoded, and hence are inappropriate to effectively manage the modern complications of managing wastes in large cities. Most of the laws deal with the general tidiness of the city streets, collection of waste resources and their disposal at places that is far away from settlements. Even these inappropriate laws are implemented entirely. This intensifies the situation further.

3.2. LEGAL FRAMEWORK APPLICABLE TO MUNICIPAL SOLID WASTE MANAGEMENT

Legislation concerning waste varies on the type of waste handling by the concerned department. International conventions are basically concerned with nuclear and hazardous waste, whereas non-hazardous waste, such as the solid waste is generally more regulated at the local or national level. From the perspective of environmental concerns, the following environmental rules, regulations, and controlling activities being implemented in various countries:

- Australia – Waste policy;
- European Union – Policies, Laws, Regulations (in English);
- Morocco – Loi relative à la protection et à la mise en valeur de l'environnement 2003 (document in French);
- South Africa – Waste Act 2008; and
- USA – Waste Legislation.

3.3. VARIOUS SOURCES OF SOLID WASTE

It is usually seen that every day, tones of solid waste get disposed of at various landfill sites all over the world. This waste primarily comes from homes, industries, offices, manufacturing sites, and various other agricultural related activities.

These landfill sites become the reason of foul smell if the waste is not treated and stored efficiently and effectively. It can result in polluting the surrounding air and could also seriously affect the health of people living nearby in addition with affecting the wildlife and our environment. Some of the common sources of solid waste are discussed follows:

3.3.1. Residential

Residences and homes are one of the major contributors of the solid place as it is the place where people live, and they contribute heavily in producing the solid waste. Garbage from these places includes food wastes, paper, plastics, glass, cardboard, leather, metals, ashes, yard wastes, and special wastes like bulky household items such as tires, electronics items, batteries, used oil, and old mattresses. Most homes have garbage bins where they throw their day-to-day solid waste and later it is collected by the garbage collecting firm or person and take to the landfill sites for treatment.

3.3.2. Industrial

Figure 3.1: Shredded solid waste.

Source: Image by Wikipedia.

Industries are considered as one of the largest contributors of solid waste, as these are the one that manufacture goods and services on a large scale. They include light and heavy manufacturing industries, fabrication plants, construction sites, power plants, canning plants, and chemical plants. These industries generate solid waste in form of food wastes, housekeeping wastes, ashes, packaging wastes, construction, and demolition materials, medical wastes, special wastes, as well as other hazardous waste.

3.3.3. Commercial

Commercial facilities and buildings are one of the other contributors of solid waste today. Commercial buildings and facilities in this scenario refer to restaurants, hotels, go downs, markets, stores, and office buildings. Some of the solid wastes produced by these comprise plastics, metals, food wastes, paper, glass, wood, special wastes, cardboard materials, and other hazardous wastes.

3.3.4. Institutional

The institutional centers such as schools, colleges, military barracks, prisons, public centers, and other government centers also have high proportion in the overall solid waste contribution. Some of the common solid wastes generated by these institution centers include glass, plastics, rubber waste, wood, food wastes, paper, metals, electronics, cardboard materials, as well as various hazardous wastes.

3.3.5. Construction and Demolition Areas

Construction sites and demolition sites also contribute heavily in the solid waste. Construction sites majorly comprise of new construction sites for buildings and roads, building renovation sites, road repair sites, and building demolition sites. Some of the solid wastes produced in these places include steel materials, cement, concrete, wood, rubber, plastics, copper wires, glass, and dirt.

3.3.6. Municipal Services

Figure 3.2: Role of municipality in collecting waste from various localities.

Source: Image by Joint Base Langley Eustis.

The urban centers also contribute massively to the solid waste crisis in many countries today. Some of the solid waste that are generated by the municipal services include, street cleaning, wastes from beaches and parks and, landscaping wastes, wastewater treatment plants, and wastes from recreational areas such as sludge.

3.3.7. Treatment Plants and Sites

Heavy and light manufacturing plants also play the role in generating solid waste. They include power plants, refineries, mineral extraction plants, processing plants, and chemicals plants. Among the wastes generated in these plants and sites include unwanted specification products, industrial process wastes, plastics, metal parts just to mention but a few.

3.3.8. Agriculture

Agriculture is one of the important sectors in the context of the economy and for providing foods to the entire population. However, agriculture produces a large amount of solid waste, for example, fruits, and vegetables that got damaged because of insects or extreme climatic conditions, etc.

This kind of problem is mostly observed in the developing countries where there are lack of proper management techniques, lack of insecticides, and limited storage and warehousing facility. Some of the solid waste are crop farms, dairies, orchards, feedlots, and vineyards are also sources of solid wastes. Among the wastes produced in the agriculture sector includes agricultural wastes, pesticide containers, spoiled food, and other hazardous materials.

3.3.9. Biomedical

This refers to hospitals and biomedical equipment as well as chemical manufacturing firms. In hospitals, there are varieties of solid wastes produced. Some of these solid wastes include bandages, syringes, used gloves, drugs, plastics, paper, food wastes, and chemicals. All these wastes require proper disposal or else it causes dissemination of infection, which affects both the people living nearby in addition to the environmental impact.

3.4. METHODS OF SOLID WASTE MANAGEMENT

Figure 3.3: Sorting of solid waste.

Source: Image by Wikimedia Commons.

There are various methods of SWM. Some of the recognized methods of SWM are discussed as follows:

3.4.1. Sanitary Landfill

This is one of the most widely used solid waste disposal methods as of today. Garbage is generally spread out thin layers, covered, or compressed with soil or plastic foam. The present-day landfills are designed in such a way that the bottom of the landfill is covered with an impervious liner that is generally made of several layers of thick sand and plastic. This liner helps in ensuring that the ground water being prevented from contaminated because of percolation or leaching. When the landfill is filled with its fuller capacity, it is covered with layers of sand, topsoil, clay, and gravel to prevent seepage of water.

3.4.2. Incineration

It is a method that is concerned with the burning of solid wastes at extreme temperatures level until the wastes are converted into ashes. Incinerators are designed in such a way that they do not produce high level of heat while burning solid wastes. This is a method of SWM that can be done by individuals themselves, municipalities or even institutions. One of the best things about this process is that it allows to reduce the volume of waste by up to 20 or 30% of the original volume.

3.4.3. Recovery and Recycling

Recovery or recycling is a process in which useful items from the solid waste are taken out through recycling or recovery techniques so that they can be used again. Basically, these items are processed in machine to clean before they are recycled. The recovery or recycling process aims at minimizing the overall energy loss, reducing the consumption of new material and reduction of waste materials in the landfill's sites.

3.4.4. Composting

Because of inadequate space in the landfill sites, biodegradable yard waste is preferable in order to decompose in a medium designed for the purpose. It is generally seen that only biodegradable waste materials are used in composting. Composting helps in the formation of good quality environmentally-friendly manure which can be used in agriculture to provide essential nutrients and minerals to the soil.

3.4.5. Pyrolysis

This is one of the methods of SWM whereby solid wastes are decomposed chemically with the help of heat without the existence of oxygen. This process is usually done under pressure and at a temperatures level of up to 430 degrees Celsius. Thus, it helps in converting solid wastes into gasses, small quantities, and a solid residue of liquid.

It is important to note that the proper SWM is an integral part of environmental conservation that should be given higher importance by both the individuals and companies at a global level. It will help in ensuring the clean environment and address the health and settlement related problems.

3.5. FUNCTIONAL ELEMENTS OF THE WASTE MANAGEMENT SYSTEM

There are six functional components of the waste management system, as discussed in the following section:

- **Waste generation**: It covers and includes all the activities and functions concerned with the identification of material which are no longer required and are either gathered for systematic disposal or thrown away.

- **Onsite handling, storage, and processing**: It includes all the activities at the point of waste generation that helps in easy management of waste resources. For example, waste bins are generally put at those sites that generate large amount of waste such as waste bins in the public areas such as markets, parks, beaches, etc. In addition, waste bins are further categorized into different types one in which liquate waste are put and other in which dry items are placed.
- **Waste collection:** Waste collection is considered as the crucial phase of waste management, as it includes the activities such as placing waste collection bins, collecting waste from those bins, and gathering trash at the sites where the collection vehicles are emptied. Although the collection phase requires the use of transportation, it is basically not the main stage of waste transportation.
- **Waste transfer and transport:** It refers to all the activities involved in movement of waste material from one place to another. These activities are majorly interested in transportation of waste material from the local generation sites to the disposal sites in large waste transport vehicles.
- **Waste processing and recovery**: It refers to all the facilities, techniques, and equipment employed to recover recyclable or reusable materials from the waste stream and to improve the overall effectiveness of other functional elements of waste management.
- **Disposal**:It refers to the final stage of waste management. It mainly covers all the activities concerned with systematic disposal of waste materials at sites such as waste-to-energy or landfills facilities.

3.6. INTEGRATED SOLID WASTE MANAGEMENT (ISWM)

As the field of SWM advances, solutions are being looked at more systematically and holistically. ISWM, for example, is an increasingly important term in the field of waste management. It refers to the selection and use of appropriate management programs, technologies, and techniques to achieve particular waste management goals and objectives.

The U.S. Environmental Protection Agency (EPA) states that ISWM is composed of waste source reduction, recycling, waste combustion, and landfills. These activities can be done in either an interactive or hierarchical way.

In closing, it is important to stress that better SWM programs are urgently needed in some countries. Only about half of the waste generated in cities and one-quarter of what is produced in rural areas is collected. Internationally, the World Bank warns that global waste could increase by 70% by 2050 in a business-as-usual scenario. Ongoing efforts to improve the waste management system are an important part of preserving a healthy human and ecological future.

Solid waste generation is a continuous growing problem at local, national, and international levels. Solid wastes refer to both organic and inorganic waste materials produced by society over a period of time through various activities, which have lost their value to the first user. A waste for one person may be consider useful for another person. If solid wastes are not disposed of properly, it can have significant negative impact on the essential elements of the living environment, (i.e., air, water, and land) at local, national, and international levels.

It is generally seen that the urban society rejects and generates solid material continuously primarily because of significant increase in production and consumption activity. The type of problem is more prevalent in the developing nations as compared to the developed ones, as the rate at which their economy grows, and the urbanization is far higher than the developing nations.

This growth and development come along with effective and efficient SWM in terms of solid waste at generation, collection, storage, processing, transfer, and transport, and disposal stages in an environmentally sound and sustainable manner in accordance with the best principles of public health, engineering, economics, aesthetics, conservation, and environmental considerations.

Thus, it can be said that the SWM is a broader concept that includes various functions such as administrative, legal, financial, engineering, and planning functions (Ramachandra, 2006; Ramachandra and Varghese, 2003).

The environmentally sound management of solid wastes issue had grabbed the attention of various national and international policy governing bodies and citizens (Subramanian, 2005). At the international level, the concerns about the solid waste generation across the world started in the year 1992 with the Rio Conference, where efficient and effective management of solid waste was considered as one of the topmost priorities.

The Johannesburg World Summit on Sustainable Development in 2002 focused on initiatives and efforts undertaken to boost the shift to sustainable consumption and production, and controlling or minimizing the resource degradation, waste, as well as pollution level.

Proper recognition has been given to waste minimization, reuse, and recycling activity followed by the safe disposal of waste at landfills site to curtail the pollution level. It is generally seen that governments of various countries have taken many initiatives and implemented new technologies and methods.

For instance, loans have been provided to these businesses that directly deals with the solid waste disposal in order to encourage for setting up composting plants for proper SWM activities (MoEF, 2005). This SWM problem has been grown rapidly with the rapid urbanization experience by various countries across the world.

With the increase in public awareness about the MSWM, people are now more conscious and ready to cooperate with municipalities in effective SWM. For instance, for the first time, after seeing the pressure from outside, government in some countries have started including private organizations in providing this public service (DPCC, 2002).

New methods of storage, collection, processing, transportation, and disposal are being explored and implemented. Every country has its own set of problems and scale of operations; therefore, it is necessary to evaluate their own pros and cons and various factors and elements before adopting and implementing transportation, he possible methods.

It requires an auditing of all functional elements of MSWM, considering the environmental constraints. An environmental audit of MSWM is generally conducted to understand the shortcomings.

Environmental auditing first began with the objective of 'polluters pay,' in order to reduce the overall liabilities towards the government. In this, various companies voluntarily participated to carried out audits of their operations and processes in order to prove that their products are environment friendly, with the increasing awareness of the public about environmental protection. It is the social and ethical responsibility of each and every company to work towards this segment in order to fulfill various goals.

It is generally seen that waste audits are undertaken for a variety of reasons, some of them are discussed as follows:

- ensure regulatory compliance;
- compare actual progress with the desired ones;
- develop baseline generation data;
- identify opportunities that works towards minimizing overall waste;
- establish sustainable development benchmarks or indicators as it helps in making the comparison easy (Ashwood et al., 1996). In general, there are three different approaches for conducting a solid waste audit, namely;
- the back-end approach, that is concerned with assessing the overall materials which is generated by the entire facility, i.e., no efforts were put to evaluate the way in which the wastes and recyclables are generated within the facility;
- the activities approach, which is responsible for tracking the overall waste generated and the waste that can be easily recyclables as they are generated throughout the facility, by performing waste audits during each process activity, e.g., an office, cafeteria or warehouse, etc.; and
- the input/output approach, which is responsible for maintaining record of material input and output related with each activity area (CCME, 1996; Dowie et al., 1998).

The primary objective of every country to introduce the environmental audit is to minimize the overall waste generation and pollution. The amendment laws or regulation imposed by the concerned department is applicable to manufacturing sector, industries, public sector, hospital or schools that have significant contribution in the waste disposal.

It is generally seen that there are many developing countries which face significant challenge in the improvement of SWM. The municipalities of many countries have taken various initiative to witness the improvement in this sector (The Expert Committee, 2000). For example, the involvement of private sector in the management of municipal solid waste has been dramatically increased. To achieve the success in the initiatives, it is of utmost importance to conduct an audit.

An audit will help in identifying and bring out the lacunae and the loopholes in the current system win accordance with the compliance with environmental regulations, resource management, occupational health, occupational health and safety, and pollution prevention systems. It is considered as the best possible way in increasing the awareness about the most suitable approaches or methods to MSWM, the issues likely to be faced and the alternative measures that can be adopted, considering the local scenario.

It is generally seen that the developed countries possess the ability to provide the required technical assistance in SWM to developing countries focusing on SWM as a technical problem with the assumption that the solid waste problem can be solved with mechanization (Lardinios and van de Klundert, 1997).

The 'blind technology transfer' of machinery from the developed nations to the developing ones and its subsequent failure has drawn the attention to look for adequate technology (Beukering et al., 1999) that can suit to the needs of the developing countries (type of waste, treatment, composition, etc.).

Composition of the waste provides a brief explanation of the constituents of the waste; this varies significantly from locations to locations as described in Table 3.3. The most noteworthy difference that is clearly observable is the difference in organic content which is significantly higher in the low-income countries as compared to the high-income countries, while the usage of paper and plastic content is much higher in high-income countries compared with the low-income ones. This shows the cultural and educational differences as well as the difference in the consumption pattern.

In higher-income countries, it is often seen that disposable material, packaged food, and magazines are used in extreme quantities, which result in having a waste of higher calorific value, lower moisture content, and lower specific density.

In the case of countries with lower income groups, the usage of fresh vegetables instead of packaged food is far higher and the materials which are reusable are mostly used. This results in a waste composition which is high in moisture content, low calorific value, and high specific weight.

Table 3.3: Relative Composition of Household Waste in Low, Medium, and High-Income Countries

Parameter (Contents physical and chemical properties)	Low-income countries	Medium-income countries	High-income countries
Organic (putrecible)	40–85	20–65	20–30
Paper, %	1–10	15–30	15–40
Plastics, %	1–5	2–6	2–10
Metal, %	1–5	1–5	3–13
Glass, %	1–10	1–10	4–10
Rubber, leather, etc., %	1–5	1–5	2–10
Other, %	15–60	15–50	2–10
Moisture content, %	40–80	40–60	5–20
Specific weight, kg/m^3	250–500	170–330	100–170
Calorific value, kcal/kg	800–1100	1000–1300	1500–2700

Source: Table from Semanticscholar.

3.6.1. Municipal Solid Waste Management (MSWM) Audit

In the past few years, auditing has become one of the most widely used tools in assessing the environmental policies, ensuring the quality of implementation, compliance with national law and regulation, etc. one of the most common used of auditing is observed in the industries and manufacturing sector.

However, in developing countries, the use of municipal solid waste audit is very rare. Whereas in western countries audits on urban waste management have been extensively carried out with respect to performance, risk, compliance, existence of waste policy, quality of implementation, etc. Most of the developed countries have set up independent organizations for assessment in the above area.

For example, the Estonian Government had carried out an audit in order to evaluate the necessary conditions for successful implementation of the waste policy. For the successful conduct of an audit, reports, and questionnaires were used and it was observed that the management had various shortcomings, such as limited finance budget, in comparison with the objective that they want to accomplish. In addition to this, the management has lack of transparency for waste management plan and also has poor monitoring and regulating policies (Linnas, 2001).

The Audit Institution of Costa Rica had carried out an audit on SWM on two municipalities, with manifold focus such as pollution prevention system, site audit, and management system. The audit was conducted and accompanied by the questionnaire, reports, interviews, and site surveys.

The various elements that were taken into consideration are compliance with national law and regulation, operational risk, occupational health and safety, resource management and pollution prevention. In the audit, it has also come under notice that there was a lack of transparency which resulted in the repetition of many working plans.

Some of the other findings were, lack of public awareness programs, inadequate new methodologies and technologies, improper monitoring, and insufficient financial support. There was also investigation about the dump with respect to the national health legislation and technical regulations. Some of the tools that were used in this audit are checklists and site surveys.

From this audit, it was observed that there was no urban cleanliness plan charted out by the municipal authority, clear violations of the legislations, no transparency and monitoring by the supervision agency and delayed closure of the dump (INTOSAI, 2002).

This audit was conducted in one of the cities of developing country, that is Bangalore in India. It will help in sharing and transfer of innovative practices that have been adopted for managing municipal solid waste.

The audit is majorly conducted by keeping in mind all the stakeholders associated with the MSWM, the current practices, the role of each entity, the shortcomings of the current practices and issues to be addressed to improve the condition. Some of the objectives of auditing of MSWM are discussed as follows:

- To review the existing MSWM practices, and
- To audit the MSWM practices.

3.6.2. Methods of Audit

The method adopted in auditing of MSWM is qualitative in nature. Information was accumulated by adopting various ways and methods to gain better insight into the situation, perspectives, issues, and priorities. Data collection methods included semi-structured interviews, document/ literature review, observation, and checklists. Audits of various different types were carried out to achieve various goals:

- **Compliance audit**. It is a type in which the current waste management process is compared with the standard set as per the legislation.
- **Operational risk audit in combination with pollution prevention audit.** It is used to check the rate at which solid waste is damaging the environment and the possible consequence of it in the coming years. The measures that have been taken against these possible environmental damages were verified.
- **Resource management audit.** To check the optimal utilization of energy, water, as well as other valuable material resources.
- **Occupational risk audit**. To verify the measures of occupational safety.

3.7. UNDERSTANDING HEALTH EFFECTS OF INCINERATION

To gain an insight onto the possible health consequences associated with the waste-incineration emissions, there is need of more information and data on contributions made by incineration to human exposures to potentially harmful pollutants and the responses that are conducive to these exposures.

There are various tools discussed in this section to understand the possible effect incineration. Of the all below discussed tools, which help in enhancing the understanding, risk assessment methods are considered as the most adequate tool in providing most-detailed information for regulatory decision makers. Some of the tools for evaluating the health effects are discussed as follows:

3.7.1. Tools for Evaluating Health Effects

Whenever assessing or analyzing for small or subtle health effects of exposures to environmental contaminants, it is suitable to consider the best possible approaches and to critically compare their results. The primary tools that have been used include risk assessments and epidemiologic studies.

These are discussed separately in the below section, although it is always advisable to make combination of approaches whenever required.

Environmental monitoring, life-cycle assessment (LCA), or biomarkers of exposure or effect, are other widely used tools that generate data which often support, confirm, or enhance the findings obtained during the conduct of epidemiologic or risk-assessment investigations. Exposure assessment is also very useful in majority of these approaches.

All these approaches are very useful in evaluating multiple environmental media (air, soil, surface water, groundwater, sediments, and any other media that could be differentiated from others), many scenarios for exposure, multiple exposure pathways, multiple routes (ingestion, inhalation, and dermal), many health end points, and multiple population groups.

However, the approaches that presently in use to examine the impact of waste incineration are typically facility-specific and site-specific, and so fail to address two important questions regarding a facility or site:

- To what extent does an incineration facility have an impact on altering the environmental concentrations of substances of concern or altering the current magnitudes of human exposure to those substances?
- What is the total contribution at local or regional level in waste incineration to human exposures?

3.7.1.1. Epidemiologic Studies

The primary objective to conduct the epidemiologic studies is to test hypotheses about the occurrence (usually incidence or prevalence) of a health outcome, to measure the strengths or sizes of relationships between such quantifiable and outcomes factors (e.g., the magnitude of exposures) or qualifiable factors (e.g., exposure status), or to generate testable hypotheses about such relationships.

The strengths and weaknesses and methodology of environmental epidemiologic studies have been discussed in previous NRC reports (NRC 1991c, 1997). Some of the key strengths of epidemiologic studies are discussed as follows:

- The people studied are likely to be those which have been exposed to the material of interest. For incinerator emissions, there is no extrapolation necessary from single chemicals to the intricate mixtures to which people are more likely to be exposed to.
- Humans themselves are being part of actual exposure conditions—there is no extrapolation from diverse animal species or diverse conditions.
- Individual and group variability in both exposure and sensitivity are inevitably taken into consideration.

3.7.1.2. Risk Assessments

Risk assessment refers to the use of procedures to evaluate the probability that harm will arise from some activities or action such as the operation of a facility. The procedures adopted to conduct risk assessments vary extensively, from a snap judgment to the use of intricate analytic models.

However, risk assessments of incineration as well as incineration facilities have become more standardized and structured.

In the case of a particular incinerator, the preliminary step, hazard identification, might started with identification of the chemicals being exists in emissions and likely to pose adverse impact on heaths. The emissions are required to be quantified, the potential health effects identified, and the scenario in which it may cause those effects defined.

The efforts undertaken to find the rough estimate of emission-rate might take the form of direct measurements, which are restricted by the sensitivity of the measuring methods, the cost of such measurements, the changeability over time of emission rates, and the inaccuracies impacting the valuable information.

After obtaining a list of chemicals identified as potentially of concern, a dose-response assessment is used in order to evaluate quantitatively the connection between exposures and toxic responses. Ideally, this assessment generally takes into consideration all the particular conditions or scenarios of exposure, including the complete mix of other potential contaminants from incineration, and exposures to the same and diverse chemicals from variety of sources.

In practice, dose-response assessments are restricted in its application, by the regulatory milieu of most risk assessments, to the use of cancer potency-slope estimates or unit risks (for the estimation of cancer risks) and reference doses (for the examination of noncancer risks) published in the Integrated Risk Information System (IRIS)[3] or other regulatory documents by the Agency for Toxic Substances and Disease Registry (ATSDR) or by the Environmental Protection Agency (EPA).

3.7.1.3. Other Tools

Biological markers of exposure or effect and environmental monitoring are two widely used tools in conjunction with epidemiologic or risk assessment investigations. These tools play very pivotal role in identifying or confirming pollutants that may likely to cause adverse health effects.

Life-cycle assessment (LCA) has been used to estimate the environmental burdens and resource consumption associated with a product, package, process, or activity throughout its entire life over large geographic regions. Life cycle assessment can be used in association with risk assessments to evaluate the effects over a broad scale—from the time when the chemicals are being introduced into the environment to its destruction phase.

3.7.1.4. Environmental Monitoring Studies

In principle, it is always viable to measure concentrations of certain pollutants directly from the incinerator in the surrounding environment. Such monitoring is most commonly of the ambient air, but soil, sediments, water, foods, and vegetation have at times been monitored for some of the emitted pollutants.

Environmental monitoring plays a very important role primarily because of the fact it straightway measures the concentrations of certain materials from a particular incinerator, in some instances in the media of immediate interest (e.g., dioxins in cows' milk or vegetation). No health effects are measured. However, as far as the assessment of health effects are concerned, environmental monitoring suffers from several disadvantages, because:

- There is generally a problem in differentiation of the contribution of the incinerator to environmental concentrations.
- Monitoring the measurements are restricted both in space as well as in time while concentrations are often highly variable in both time and space.

Because of reasons, environmental monitoring is considered as most relevant in calibrating, confirming, or disproving the modeling efforts used in risk-assessment methodology.

3.8. HAZARDOUS SUBSTANCES ASSOCIATED WITH WASTE MANAGEMENT

Monitoring of environment in order to gain an insight into the potential sources of pollution from different waste management options has been and is being consistently carried out and thus a great deal is known about the different kinds and quantity of substances releasing from them.

Whatever the option of waste management is, it is generally the case that: (a) there are generally extreme amount of a different substances; and

(b) only a few of these are produced in any quantity with many being at extremely low levels.

Gases which are released from the landfill sites, for instance, consist majorly of carbon dioxide and methane, with other gases, such as mercury vapor and hydrogen sulfide being emitted at low concentrations, and a mixture of volatile organic compounds (VOCs) comprising approximately 0.5%.

A WHO exposure assessment expert group proposed that priority pollutants should be stated on the basis of environmental persistence, toxicity, and mobility, bioaccumulation, and other hazards such as explosivity.

In addition to the above-discussed substances, they are also of the opinion that the investigation of the landfill site should consider metals, polychlorinated biphenyls (PCB), polycyclic aromatic hydrocarbons (PAH), pesticides, chlorinated hydrocarbons, dioxins, pharmaceuticals, asbestos, and pathogens.

Waste incineration is also responsible for generating a large number of pollutants from the combustion of clinical, chemical, and municipal waste, and sewage sludge, which can be grouped as particles and gases, organic, and metals compounds.

There are generally ten pollutants that believed to have a significant adverse impact on human health based on environmental persistence, bioaccumulation, and amount emitted and/or on inherent toxicity were mercury, cadmium, arsenic, nickel, chromium, dioxins, PAHs, PCBs, PM_{10} and SO_2. Microbial pathogens are considered as a potential source of hazard, especially in sewage and composting treatment but also in a landfill. Dust and the production of particulate matter are released in the incineration, landfill, and composting processes and by road traffic included in all waste management options.

Less easily quantifiable hazards, which might nevertheless have an effect on the population near a waste disposal site, include odors, litter, noise, heavy traffic, flies, and birds.

3.9. SUSTAINABLE SOLID WASTE MANAGEMENT

Figure 3.4: Sustainable sanitation.

Source: Image by Wikimedia Commons.

Sustainable SWM can be defined as the technique or methods aimed at reducing the overall amount of natural resources consumed, ensuring that any materials that are extracted from nature are reused as many times as possible and that the waste created is kept to a minimum.

The primary objective to engage in sustainable SWM is to address concerns related to environmental pollution, public health, resource management, land use, and socio-economic impacts associated with improper disposal of waste.

According to the former US President Jimmy Carter:

> **"This growing mountain of garbage and trash represents not only an attitude of indifference toward valuable natural resources, but also a serious economic and public health problem".**

These words from are enough to understand the social, economic, and environmental impact of mismanaged waste disposal and an urgent call for help to look for smart, innovative, sustainable as well as effective waste disposal techniques.

3.9.1. Benefits of Sustainable Solid Waste Management Systems

It is generally seen that the good waste management systems always attain the maximum practical benefits from products and at the same time generate the minimal amount of waste. Effective waste management systems always help in getting maximum co-benefits beyond addressing emissions, which might consider as the primary function of the waste management. Some of these co-benefits include:

Public health: Improper management of waste often ends up in the streets or water drains, attracting pests and vermin, crating room for bacteria or insects. Waste disposed in dumps or unsanitary landfills can contaminate underground water with toxic leachate.

Improving the overall waste collection and maintaining effective disposal practices can have a direct impact on the quality life of people, improved health conditions, access to clean water and a cleaner city environment.

Air quality: The open burning of waste is a common practice followed by many countries across the world and a major source of black carbon. Improper planning of waste collection routes or the use of old vehicles for waste collection can significantly increase the amount of emission released in the aim, negatively affecting air quality. Implementing the smart strategy for collection of material and looking for alternatives of burning waste such as dumping the waste into landfills instead of burning help in improving the air quality.

Poverty reduction: It is often seen that in many cities, waste is a significant source of income for large number of people and act as a raw material for many industries across the world. Waste collectors in many cities operate on the streets and dumps, include a wide range of activities range from collecting, sorting, recycling, cleaning, to selling materials thrown away by others.

Various initiatives and action taken by the cities' municipalities can have a significant impact on the economic conditions and overall quality of life of those people, business or organizations that are directly associated with the waste management. For example, by utilizing proper sanitary landfill disposal techniques, cities can avoid the health hazards that could be the result of open dump scavenging.

Social justice: Solid waste management is highly visible and affects attitude of people towards government and society itself. Ensuring effective planning and sustainable investments in MSWM systems necessitates comprehensive understanding of the needs and preferences of a wide range of stakeholders associated with service delivery, costs, and corresponding environmental and social impacts.

3.9.2. Challenges to Delivering Sustainable Waste Management Systems

It is generally seen that within the past few years, SWM has been a consistent challenge for city authorities, primarily because of the significant rise in the generation of waste, the burden it places on municipal budgets, the limited information about the number of factors that have an impact on the overall SWM and of the necessary linkages to enable effective function of the entire handling system. one of the most major challenge that need to be addressed is to implement a successful SWM system.

Some of the challenges that are faced by the concerned department in SWM all around the world are discussed as follows:

Complexity of waste management: SWM is a multi-dimensional issue and there are various stakeholders that are associated with its that engages multiple stakeholders. Municipalities in general seek equipment to find solutions to the diversity of problems they face. But a successful waste management system must consider technological solutions along with environmental, socio-cultural, legal, institutional, and economic linkages. It also needs to address syndromes like NIMBY (not-in-my-backyard) that can be prevalent amongst the public.

Involvement of multiple stakeholders: Waste management involves many stakeholders with different and sometimes conflicting interests. A detailed understanding of whom the stakeholders are and the responsibilities they have in the waste management structure is required to establish an efficient and effective system. It is very important to engage in the effective communication among the stakeholders in order to ensure a well-functioning waste management system, especially in the developing country's' cities.

Institutional challenges: Many municipalities, majorly the solid waste departments, have a shortage of staff and the staff that they have do not possess the relevant skills level.

This lack of human resources to deal with enormous solid waste often result in unconventional methods of disposal, which include open dumping and burning. It necessitates the need on part of municipalities to do adequate investment in capacity building, both in terms of quality and number of staff resources and skills.

Difficulty in recovering costs: There are number of direct and indirect cost that is associated with solid waste services, which is very difficult for municipal government to recover in a given period of time. Because of inadequate return on investment, they resist from doing investment in this area. They lack the monetary resources which is important in hiring the experienced personnel, in acquiring the equipment and infrastructure needed to implement waste management plans.

Implications beyond municipal boundaries: It is important to have access of authentic and reliable data in order to create proper information channels within and between municipalities about waste management. The person that are responsible for key decisions should be well aware and informed about the situation of the cities in order to develop the plan and implement strategies accordingly. It helps in making the positive changes, developing integrated waste management strategies adapted to the needs of the citizens.

3.10. CONCLUSION

In the end, it is concluded that solid waste auditing is very important in order to know the actual picture of the current scenario to plan and strategies accordingly. With the increase in the population across the world, it becomes difficult to deal with the amount of waste that is generated worldwide.

For proper management of solid waste, it is foremost important to, first of all, know about the various sources of waste. After then, the decisions related to which waste can be recycled or reuse can be taken.

In addition, there are various methods of SWM, selecting the best method is one of the critical tasks in achieving the objective of SWM. Industries should voluntarily participate in sustainable SWM in order to reduce the amount of total waste generated.

REFERENCES

1. academia.oup.com. (2003). *Health hazards and waste management*. [online] Available at: https://academic.oup.com/bmb/article/68/1/183/421368 (accessed on 4 March 2020).
2. Ahsan, A., Alamgir, M., El-Sergany, M., Shams, S., Rowshon, M., & Daud, N., (2014). Assessment of municipal solid waste management system in a developing country. *Chinese Journal of Engineering*, (pp. 1–11).
3. Conserve Energy Future, (n.d.). *What is Solid Waste Management? Sources and Methods of Solid Waste Management-Conserve Energy Future*. [online] Available at: https://www.conserve-energy-future.com/sources-effects-methods-of-solid-waste-management.php (accessed on 4 March 2020).
4. Environmental Audit of Municipal Solid Waste Management, (n.d.). [eBook] Available at: http://wgbis.ces.iisc.ernet.in/biodiversity/pubs/ces_tr/TR118_SPoonancha/Shruthi_report_12_TVR_r_230407.pdf (accessed on 4 March 2020).
5. Environmental-auditing.org. (n.d.). *Performance Audit Report on Solid Waste Management*. [online] Available at: https://www.environmental-auditing.org/media/4675/samoa_f_eng_solid-waste-management.pdf (accessed on 4 March 2020).
6. https://pdfs.semanticscholar.org/d990/bf3c41b519e9df5401165a0e9f7b2119f9e5.pdf. (2007). [eBook] Available at: https://pdfs.semanticscholar.org/d990/bf3c41b519e9df5401165a0e9f7b2119f9e5.pdf (accessed on 4 March 2020).
7. Jassim, A., 2017. Sustainable Solid Waste Recycling. *Skills Development for Sustainable Manufacturing*, [online] Available at: <https://www.intechopen.com/books/skills-development-for-sustainable-manufacturing/sustainable-solid-waste-recycling> [Accessed 4 March 2020].
8. Muraleedharan, S., (2019). *Sustainable Solid Waste Management: Need of the Hour*. [online] BioEnergy Consult. Available at: https://www.bioenergyconsult.com/sustainable-solid-waste-management/ (accessed on 4 March 2020).
9. Ncbi.nlm.nih.gov. (2020). *Understanding Health Effects of Incineration*. [online] Available at: https://www.ncbi.nlm.nih.gov/

books/NBK233619/ (accessed on 4 March 2020).

10. Performance Audit of Municipal Solid Waste Management, (2015). [eBook] Available at: https://www.sao.ge/files/auditi/efeqtianobis-angarishi/Solid-Waste.pdf (accessed on 4 March 2020).
11. Ppp.worldbank.org. (2020). *Solid Waste Laws and Regulations | Public Private Partnership*. [online] Available at: https://ppp.worldbank.org/public-private-partnership/sector/solid-waste/waste-laws-regulations (accessed on 4 March 2020).
12. Solid Waste Management Assessment, (2020). [eBook] Available at: https://www.epd.gov.hk/eia/register/report/eiareport/eia_0432000/theme/07.pdf (accessed on 4 March 2020).
13. Sustainable Solid Waste Systems, (n.d.). [eBook] Available at: https://c40-production-images.s3.amazonaws.com/good_practice_briefings/images/9_C40_GPG_SSWS.original.pdf?1456789082 (accessed on 4 March 2020).
14. UNEP-UN Environment Programme. (2020). *Solid Waste Management*. [online] Available at: https://www.unenvironment.org/explore-topics/resource-efficiency/what-we-do/cities/solid-waste-management (accessed on 4 March 2020).
15. Waste Audit, (n.d.). *What is a Waste Audit?* [online] Available at: https://www.wasteaudit.com.au/wp_super_faq/what-is-a-waste-audit/ (accessed on 4 March 2020).
16. Waste management—audit guidelines, (2020). [eBook] Available at: http://iced.cag.gov.in/wp-content/uploads/2013/09/Waste%20audit%20guidelines.pdf (accessed on 4 March 2020).

CHAPTER 4

Air Quality Impact Assessment

CONTENTS

The chapter of air quality impact assessment explains the basic significance of air quality impact assessment. This chapter also explains the application of the statistical tools in order to assess the air pollution. This chapter also includes the materials and methods that are used for assessing the air quality impact such as sample points, multivariate analysis, and spatial analysis. This chapter also mentions the air quality regulations and the assessment criteria.

This chapter also addresses the impact assessment which includes, atmospheric emission inventory, and stack parameters and emissions. This chapter provides highlights on the factors modifying source and/or emissions such as regulatory context, demand, urban planning, and behavior. This chapter also explains factors modifying concentration such as topography and meteorology, green spaces, and architecture.

4.1. INTRODUCTION

Air quality impact assessment (AQIA) is considered as an important technique. This technique is used for deciding the relative contribution to ground-level pollutant concentrations of particular current or future source emissions at receptor sites. Air quality modeling and monitoring techniques are the main activities in AQIA.

The selection of the techniques which are applicable to a specific situation is intimately associated to the problem that are to be assessed. The review of a relevant modeling techniques has been introduced. The main focus of this technique is placed on Gaussian plume models because of their ease of use and broad applicability.

There are several alternative modeling techniques that have been suggested when departures from the Gaussian form in the atmosphere takes place or when more detailed information regarding atmospheric chemistry, deposition or long-range transport is needed.

The purposes and techniques of ambient air quality monitoring have been analyzed and the sitting of duration and frequency of sampling, instruments, selection of monitoring rationale (that is fixed or mobile), and reduction of sampling errors, data storage, and analysis are discussed. For successful AQIA, modeling, and monitoring are very important.

An orderly approach to the principles and procedures that is included in performing AQIA studies, by the use of these two techniques in conjunction, has been presented.

Figure 4.1: Impact of pollution to the air quality.

Source: Image by Needpix.com.

4.2. WHAT IS AIR QUALITY IMPACT ASSESSMENT?

In general terms, an air quality assessment must be conducted if there is a new development, increment in the usage of an existing property, or alterations in the usage that will likely bring about significant changes.

This may involve increases in traffic movements, or the initiation of a new emission. This new emission may have impact on the local air quality, like a biomass boiler that may be within an Air Quality Management Area (AQMA). In the absence of correct assessment and awareness of air quality sources and their associated impact to identified receptors it would not be possible to propose efficient mitigation solutions.

Thus, it has been observed that the demand by local planning authorities for air quality assessments has increased in a significant manner. It has increased in support of planning applications for new developments or change of use proposals, where there are already sensitive receptors near, or the site is within an AQMA.

Figure 4.2: Basic significance of air quality assessment.

Source: Image by Pxhere.

Further, this is defined in the National Planning Policy Framework (NPPF). It states that 'Planning policies should sustain compliance with, and it should contribute towards EU limit values or national objectives for pollutants, considering the presence of AQMAs and the cumulative influences on the quality of the air from individual sites in local areas.

All the planning decisions should make sure that any new development in AQMAs is consistent with the local air quality action plan.'

Thus, an Air Quality Assessment should be considered where:

- Increase in the deposition of nitrogen or a proposal for a new development or a new source of exposure that is located closely to the existing sources (such as a busy road).
- A development proposal will make a significant change in traffic volumes (like a change in annual average daily traffic (AADT) and peak flows), congestion, vehicle speeds, parking, and/or traffic composition (like increases in HGV because of the commercial use).
- A new emission source will be presented like a biomass boiler or CHP burning plant within AQMA or in close proximity to an AQMA.

- A sizable construction site is proposed as part of a new development. This will make significant HGV movements each day (such as >200) over a given period, often it is 1 year.
- Proposals could have a potential influence on all the nearby sensitive receptors like emissions impacting on residential housing, formation of dust from construction works and many other things.

The need for the points that are mentioned above will totally depend on the geographical location and the sensitivity of the proposed source. Also, other factors such as the requirements of local planning authority, their policies and the action plans that are taken into consideration.

4.3. STATISTICAL TOOLS FOR AIR POLLUTION ASSESSMENT

It has been observed that during the last year, urban air pollution concentrations have increased all over the world. According to the World Health Organization (WHO), this increase in the urban air pollution concentrations can be estimated at 8% from the year 2008 to year 2013 and over 80% of individuals living in urban areas, where air pollution is monitored, are exposed to such levels that exceed the limits. This limit is given by WHO.

Globally, urban air pollution is considered as a serious environmental problem. As the quality of urban air declines, the risk of stroke, lung cancer, heart diseases, and chronic and acute respiratory diseases, this includes asthma, increases. In addition to it, it also contributes to damaging building materials as well as the cultural objects.

All the harmful effects of air pollution and all of its causes are extensively studied and, the decline in the urban quality are particularly associated to the increase in traffic emissions, transport-related emissions being the major component of air pollution.

It has been observed that a wide range of air pollutants are released by vehicles that have a petrol-derivatives engines, being the most important of them; carbon monoxide (CO), nitrogen oxides, volatile organic compounds (VOCs), and particulate matter have a significant influence on the quality of air in the urban areas.

It has been observed that air pollution in big cities and close to the main roadways is dominated by traffic on the road. But the level of the pollution is very variable. This is because air pollution is impacted severely by various environmental or meteorological factors. In addition to it, it is also impacted by the traffic patterns, orientation, and size of buildings or land use.

Figure 4.3: Statistical tools for air pollution assessment.

Source: Image by Pxhere.

Therefore, finding out the population exposures is very important in order to study and understand the causes of these variations. It is important to understand the variations prior to the development of interventions and policy recommendations aiming at reduction exposures. In this sense, multivariate statistical techniques are considered as an excellent tool. This tool is used to discover and analyze a large dataset of environmental data.

There are various methods of dealing with this extensive amount of data. One of the most interesting methods to treat the extensive data is by means of the application of multivariate analysis methods (that is the principal component analysis or cluster analysis(CA)).

The primary objective is aimed at grouping as well as classification of objects (in this case, stations, measured parameters, days, etc.). In addition to it, it also aims in modeling relationships between the different environmental data.

It has been observed that the methods of multidimensional analysis have made it possible to develop some correlations between different parameters. At the same time, finding correlations between the amounts of several pollutants is also made possible. There are various multivariate methods that can be used in environmental studies. This is because these methods provide information regarding the interpretation, association, and modeling from large environmental datasets.

Correlation analysis is defined as a very useful statistical tool. This tool is used to identify the relationship between pollutants or other variables that have impact on the quality of the air. Also, this method is very useful to understand or look for the most influential factors or sources of chemical components.

This statistical tool has been used in various studies which were done on air pollution in Chicago. It also performs isotopic analyses in topsoil and recognizes all the sources and correlations between PAHs and heavy metals in Switzerland and Spain and also, in the urban road dust of Xi'an (China).

Principal component analysis (PCA) is similar to many of the multivariate methods of analysis. This analysis is based on data reduction. The correlation between the data is considered in PCA. This is possible because in a dataset, only a small number of parameters are significant.

PCA has been used widely in environmental analysis. This is because this statistical analysis method is proved to be a very helpful aid in data interpretation and classification. In Particular, PCA has been used collectively with the other multivariate techniques, like canonical correlation analysis (CCA). CCA is helpful in uncovering the existing relationships between meteorology and air pollutants concentrations. PCA has also been used together with CA.

The aforementioned CA, or it can be called as a more correctly hierarchical CA. It is a sorting method that is used to divide the data that is present in clusters. With the help of this method, the objects are aggregated stepwise in accordance to the similarity of their features.

As a result of this, there is the formation of hierarchically or non-hierarchically ordered clusters. It has to be noted that the ideal number of clusters may be determined graphically through a dendrogram. In general, it can be said that CA is defined as a useful procedure that is used for simplifying and classifying the behavior of environmental pollutants in a particular region.

Spatial interpolation method is the other commonly used approach in air pollution studies. Often, spatially continuous data of environmental variables are needed for environmental sciences and management. Though, information that is required for environmental variables is generally collected by point sampling. This is mainly for the mountainous region and deep ocean area.

Therefore, methods that are generating such spatially continuous data by the use of point sampling become the important tools. However, spatial interpolation methods are often data-specific or even variable-specific. There are several factors that have impact on the predictive performance of the methods. Also, there were many studies done that have shown that the effects of these factors are not consistent.

Therefore, for a given dataset, it is difficult to choose a correct method. Among all the spatial interpolation approaches, a geostatistical method like ordinary kriging (OK) is the method that is used most widely. This is because they provide concentration estimates. These estimates are provided together with standard error at unmonitored locations.

It is possible to detect a large number of air pollutants in ever-decreasing concentrations. This is possible with increasing improvement in the analytical equipment and the development of new techniques. On collecting all the information on the quality of air from monitoring stations, instead of having a restricted number of sampling points, it is possible to extract useful information by the use of multivariate methods. Those methods are used that are sufficient for the full description of a given scenario or any situation.

In this, the application of statistical methodologies like multivariate analysis with principal component analysis (PCA) and CA or geostatistical analysis with ordinary kriging (OK) could be a very beneficial aid in their interpretation of air pollution data.

So, although Madrid and its suburbs constitute one of the largest metropolitan areas in Europe, there are fewer studies that have been published about the air quality in this area.

Therefore, the main purpose is to study and examine the air quality of a city by taking the example of Madrid (Spain), its metropolitan area, along with the other areas and regions on the basis of the levels of certain compounds in the air. Major compounds that are to be assessed are nitrogen oxides, ozone (O_3), and particulate matter (PM10). Also, the distribution of these pollutants is mapped which helps in recognizing and understanding the relationship between the pollutants and the demography of that particular region.

Correlation analysis, CA, PCA, and geostatistical methodologies have been employed for this propose.

4.4. MATERIALS AND METHODS

4.4.1. Sample Points

The Madrid Autonomous Region, which is officially known as the Community of Madrid, is situated in the center of the Iberian Peninsula. At present, there are over 6.5 million individuals who are living in the region. Among all these individuals, near about 85% of individuals live in the Madrid city and its metropolitan area, being the largest in Spain. It is the 4th largest in the European Union, and also, the 54th largest in the world.

It has to be noted that the largest suburbs are in the south, and in general, along the main routes leading out of the Madrid city. The levels of pollution have been decreasing in recent years because of the pollution control policy. But this level of pollution can once again worsen as a result of the end of the economic recession and the increase of the consumption of fossil fuel.

The emission in Madrid region involves typical diffuse urban sources and industrial sources. In addition to it, it also includes traffic emissions because of intense traffic across the ring road, main urban routes and major highways that are crossing the region.

The Community of Madrid has two monitoring networks for the control of quality of the air. It has air pollution control stations which are distributed along the region, one is addressed directly by the Community government, created by 23 automatic stations (13 urban, 4 suburbans, and 6 rural stations). The other network is controlled by the municipal authorities of the city of Madrid, with another 24 stations (21 urban and 3 suburban stations).

Pollutants like NO_2, SO_2, PM_{10}, VOCs, NO, CO, and O_3 are measured continuously by reference methods at these stations. In order to date, just four of the pollutants NO_2, NO, PM_{10}, and O_3 are measured at the same time, at 22 monitoring stations being those the stations that are used for this assessment.

4.4.2. Multivariate Analysis

The statistical methods that are used in this study were Pearson's correlation coefficients, PCA, and hierarchical CA. The correlation coefficient is a statistic tool that is used to calculate the extent of the relationship between variables, when it is compared in pairs.

There are various kinds of correlation coefficients: Pearson's correlation is a correlation coefficient. This is commonly used in linear regression. Also, it has been used to evaluate the strength of relationships between the four air pollutants.

PCA is defined as a multivariate technique that is used to decrease the dimensionality of a dataset (preferably normally distributed, but not necessarily) that contains many interrelated variables. The dataset is transformed into a new set of independent (that is uncorrelated) variables or principal components (PCs).

It has been observed that most of the variation in the dataset can be explained by considering a small number of PCs when there is a significant correlation between variables. It becomes possible to visualize the patterns and correlations that are present between the data along with retaining maximum amount of information that is present in the original set of data.

It has to be noted that PCs are the eigenvectors of a covariance matrix or a correlation matrix. Every single PCs extracts a maximal share of the total variance. It has to be noted that a PCs which is having an eigenvalue greater or equal to 1 is considered as being of statistical significance (Kaiser criterion).

CA is a beneficial procedure that is used, as classification tool, very convenient in environmental studies. This is because it simplifies and complements the PCA. This method (CA) was undertaken following the Ward algorithmic method. It maximizes the variance between groups and also, reduces it between members of the similar group, and the Euclidean distance to calculate the measurements.

4.4.3. Spatial Analysis

Spatial analysis or spatial statistics was performed in order to study geometric, topological, or geographic properties. It begins from the principle that there is a relationship between the distribution of analyses and that it is very probable that the values that are attained from nearby points are more similar to each other as compared to the values from more distant points.

Spatial analysis consists of a variety of techniques, many of these techniques are still in their early development, that are making use of different analytic approaches, like spatial interpolation. This is also known as Geostatistics. It is defined as the set of tools that are needed in order to analyze spatial patterns and estimate the values of a continuous variable that are distributed in space or in time at unsampled locations.

OK is one of these methods. This method makes use of only sample data for interpolation of the missing values. It is also considered as the best linear unbiased estimator (BLUE). Geostatistics tools are used in order to determine air pollution concentrations. This allows elaborating contour maps for a more visual assessment of the distribution of a certain pollutant.

4.5. AIR QUALITY REGULATIONS AND ASSESSMENT CRITERIA

Prior to evaluating the influence of current as well as proposed activities on the atmospheric environment, reference is required to be made to guidelines governing emissions, environmental regulations and impacts of such operations.

Air quality guidelines (AQGs) and standards are very important to efficient air quality management. This provides the link between the user of that air at the downstream receptor site and the source of atmospheric emissions. The ambient air quality standards as well as guideline values show safe daily exposure levels for most of the population. This population includes the very young and the elderly, throughout the lifetime of an individual. Usually, the AQGs and standards are given for specific averaging or exposure periods.

Figure 4.4: Air quality regulation and assessment criteria.

Source: Image by Pixabay.

This section gives a brief about both the national and international legislation relating to air pollution and criteria pollutants that are relevant to the present study. A discussion on inhalation health risk and cancer risk that are related with non-criteria pollutants is also provided.

4.5.1. NATIONAL REGULATIONS

Namibia has an Atmospheric Pollution Prevention Ordinance (NAPPO) (No. 11 of 1976). This addresses the following:

- Part II – Control of gases that are toxic or offensive.
- Part III – Atmospheric pollution by smoke.
- Part IV – Dust control.
- Part V – Air pollution by fumes that are released by vehicles.

However, The Namibian Atmospheric Pollution Prevention Ordinance (No. 11 of 1976) does not involve any ambient air standards, and it just recommends opacity guidelines for smoke under Part I. It is inferred that the Director1 provides AQGs for consideration at the time of issuing of Registration Certificates.

Registration Certificates are just published for so-called "Scheduled Processes" which are processes that are resulting in toxic or offensive gasses. Generally, these processes pertain to point source emissions. Scheduled Processes at DPMT Tsumeb involves Sulfuric Acid Processes and Copper

Processes. In addition to it, arsenic Processes would also have required to be considered up to the end of the year 2016,

A range of pollutants is defined by the Ordinance as noxious or toxic and offensive gasses, but generally, the air pollution guidelines are mainly for criteria pollutants. These are names as oxides of nitrogen (NOx), SO_2, ozone (O_3), carbon monoxide (CO), lead (Pb) and PM.

4.5.2. International Guidelines and Regulations for Criteria Pollutants

Reference is made to international criteria when no local ambient air quality criteria exist or are in the process of being developed. This provides a sign of the severity of the potential impacts from the proposed activities.

The most broadly referenced international air quality criteria are those that are published by the World Health Organization (WHO).

In the year 1987, AQGs have been published by the WHO. And these guidelines were revised in the year 1997.

Since the finalization of the second edition of the AQGs for Europe which consists new research from low-and middle-income countries where the level of air pollution are at their highest, the WHO has undertaken to analyze all the collected scientific evidence and to consider its implications for its AQGs.

The result of this entire work is shown in detail in 'AQGs, Global Update 2005.' This work is shown in the form of revised guideline values for selected criteria air pollutants. These are applicable all over the WHO regions.

It was given that the level of air pollution in developing countries frequently far exceed the suggested WHO AQGs, interim target (that is IT) levels were included in the update. These are higher as compared to that of the WHO AQGs themselves, to encourage steady progress towards meeting the WHO AQGs.

It has been observed that there are between two and three ITs starting at WHO interim target-1 (IT-1) and IT-2 or IT-3 before reaching the AQGs. Interim target-1 (IT-1) is the most lenient. IT-2 and IT-3 are more stringent targets. The WHO allows a 1% frequency of exceedance per calendar year.

There are several references that have been made to the European Commission Air Quality Legislation. These references were published in the EC Directives 2008/50/EC, 1999/30/EC and 2004/107/EC. In the similar manner, the limit values were published by the European Commission. These limit values are almost similar to the values mentioned by SA NAAQS. All these values are clearly explained in the table below.

The South African Bureau of Standards (SABS) was involved to help the Department of Environmental Affairs (DEA, then called the Department of Environmental Affairs and Tourism or DEAT)in the facilitation of the development of ambient air quality standards from the year 2004.

In addition to it, this consists of the establishment of a technical committee in order to oversee the development of standards. Standards were determined which was based on international best practices for PM10, dust fall, PM2.5, NO_2, SO_2, O3, CO, Pb, and benzene.

Table 4.1: Assessment Guidelines and Standards for Criteria Pollutants Considered in the Assessment

Pollutant	Averaging (exposure period)	Limit value ($\mu g/m^3$)	Limit value (ppb)	Limit value reference	Permitted frequency of exceedance	Reference
PM_{10}	24-hour	75	Not Applicable	WHO IT-3 and SA NAAQS	4	WHO and SA NAAQS
	1-year	40	Not Applicable	SA NAAQS	-	-
$PM_{2.5}$	24-hour	40	Not Applicable	WHO IT-3 (a) and SA NAAQS	4	WHO and SA NAAQS
	1-year	20	Not Applicable	EU, WHO IT-2 (b) and SA NAAQS	-	-
SO_2	10-minute	500	191	SA NAAQS	526	SA NAAQS
	1-hour	350	134	EU, SA NAAQS	24, 88	EU, SA NAAQS
	24-hour	125	48	EU, WHO IT-1 and SA NAAQS	3, 4	WHO, EU, and SA NAAQS
	1-year	50	19	SA NAAQS	-	-

Source: https://s21.q4cdn.com/589145389/files/doc_documents/EN/ESIA/Appendix-F-Air-Quality-Impact-Assessment.pdf.

4.6. IMPACT ASSESSMENT

4.6.1. Atmospheric Emissions Inventory

The creation of an emissions inventory, which is comprehensive in nature, formed the basis for the assessment of the air quality impacts in the area of Tsumeb smelter complex. This emission inventory proposed certain upgrades for the environment.

Sources of emission and associated pollutants that are considered in the emissions inventory include:

- **Stack emissions:**
  - i. The copper stack is included, emissions from the hygiene and new converter baghouses (SO_2, $PM_{2.5}$, PM_{10}, and arsenic);
  - ii. The new RHF baghouse stack (PM_{10}, SO_2, $PM_{2.5}$, and arsenic);
  - iii. The acid plant stack (SO_2 and H_2SO_4)
- **Fugitive emissions (SO_2, PM_{10}, $PM_{2.5}$, and arsenic):**
  - i. Furnace and converter fugitives: they are consisting of:
    - a. Converter building area;
    - b. Ausmelt building area;
    - c. New RHF building area;
  - ii. Materials handling and transfer (PM2.5, PM10, and arsenic);
  - iii. Vehicle entrained dust due to product, raw materials, and waste transport on internal paved and unpaved roads (PM2.5, PM10, and arsenic);
  - iv. Crushing and screening (PM10, PM2.5, and arsenic) and
  - v. Windblown dust from raw materials areas, waste, and tailings disposal areas.

Figure 4.5: Impact assessment: Atmospheric emission inventory.

Source: Image by Piqsels.

It has been observed that emissions were measured for both the current or "base" scenario and a design or "Project" scenario. Here, the base scenario is representative of current activities at a current concentrate processing rate of 240 000 t/a. On the other hand, the project scenario comprises proposed plant upgrades with a concentrate processing rate of 370 000 t/a.

4.6.2. Stack Parameters and Emissions

4.6.2.1. Current Point Sources

Emissions that were emitted to the atmosphere from point sources at DPMT were estimated. This was based on isokinetic stack sampling that was conducted by Skyside in January 2018. Point sources at DPMT comprise the Copper Stack, the Ausmelt stack and the Acid Plant stack.

Through copper stack, the captured secondary emissions from the Ausmelt furnace and converters are emitted. The Ausmelt stack is used as a by-pass for primary Ausmelt furnace emissions when the acid plant is off-line. And through the Acid Plant stack, the tail gas from the acid plant is vented to the atmosphere.

Stack parameters (as it was reported in the earlier studies) and the emission rates (as recorded by Skyside in January 2018) are shown in Table 4.2. The following assumptions were made:

- The secondary emissions, which were captured from the converters, was recorded by Skyside before the baghouse. It was assumed that the concentration of SO_2 remained unchanged through the baghouse. And the concentration of arsenic and that particulate are removed with 90% efficiency in the baghouse.
- The yearly total of emissions through the Ausmelt stack were estimated. This was based on the availability of acid plant and utilization that was provided by DPMT. Dispersion modeling simulations were carried out for 75% utilization and for 90% utilization scenarios. It has to be noted that only the primary off-gas from the Ausmelt furnace is emitted through this stack when the acid plant is off-line, as the converters go through an emergency roll-out when the acid plant trips.
- The results of emission on the secondary emissions from converters was captured at a time when the emissions were above average (while at that time, the converters were blowing). In the

year 2016, Skyside performed a sampling campaign before the baghouse serving the secondary converter emissions which is considered to be much more representative of secondary converter emissions. This sampling campaign was carried out for 7 hours.

Table 4.2: Stack Parameters and Emissions -Current Operations

	Copper Stack	**Acid Plant Stack**	**Ausmelt Stack**	
			(75% Acid Plant Utilization)	**(90% Acid Plant Utilization)**
Average volumetric flow rate (dNm^3/h)	259 163	101 150	21 512	8 602
Diameter (m)	3.05	1.61	3.962	3.962
Height (m)	137	80	145	145
Temperature (°C)	58.4	57	76.5	76.5
SO_2 Emission Rate (t/a)	2799.4	467.0	32768	13107
PM_{10} Emission Rate (t/a)	20.0	0	50.6	20.2
$PM_{2.5}$ Emission Rate (t/a)	20.0	0	50.6	20.2
Arsenic Emission Rate (t/a)	2.3	0.0	6.7	2.7

Source: https://s21.q4cdn.com/589145389/files/doc_documents/EN/ESIA/Appendix-F-Air-Quality-Impact-Assessment.pdf

4.6.3. New RHF Baghouse Stack

Volumetric gas flow rates, emissions, and temperatures for the several operational cycles of the proposed RHF, were supplied by DPMT. It was supplied as determined by the furnace design.

A minimum stack height of 70 m was previously determined from dispersion simulations. The guideline of the IFC was adopted in the process of choosing a suitable impact criterion. It was recommended by IFC in their General Environmental, Health, and Safety (EHS) Guidelines, that the facilities should be designed in a way that it meets 25% of impact criteria to permit for the impact of other sources of emission.

Emissions and parameters for the new RHF baghouse stack are contained in Table 4.3.

Table 4.3: RHF Cycle Design Emission Rates, Volumetric Flow Rates, and Temperatures as Supplied by DPMT

Cycle	RHF charging	Matte pouring	Slag pouring	One-fan charging
Hours in 24-hour cycle	3.6 (15%)	2.5 (10%)	2.9 (12%)	0.0 (0%)
Dry standard volumetric flow rate (dNm^3 /h)	231 450	168 450	168 450	115 950
Actual volumetric flow rate (m^3/h)	335 031	237 423	236 526	182 708
Off-gas temperature (°C)	54.0	46.0	44.0	83.0
SO_2 concentration (mg/dNm^3)	1 075	1 725	87.5	2 143
PM concentration (mg/dNm^3)	0.045	0.086	0.087	0.163
As concentration (mg/dNm^3)	0.022	0.043	0.043	0.082
SO_2 emission rate (g/s)	69.1	80.7	4.1	69.0
PM10 emission rate (g/s)	0.003	0.004	0.004	0.005
PM2.5 emission rate (g/s)	0.003	0.004	0.004	0.005
As emission rate (g/s)	0.001	0.002	0.002	0.003

4.7. FACTORS MODIFYING SOURCES AND/OR EMISSIONS

To the extent that the origin of air pollution' are mostly man-made, the factors that have impact on the sources and emissions of air pollution are also associated to how all the individuals and institutions govern the use of natural resources, technology, and space.

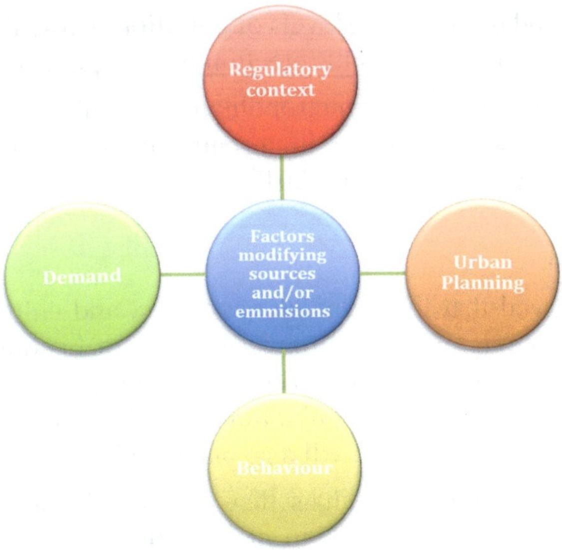

Figure 4.6: Factors modifying sources and/or emissions.

All these factors are categorized in terms of demand, regulations, urban planning, and behaviors, but it is vital to note that all the categories are flexible and the factors that within different categories can have an impact on each other.

4.7.1. Regulatory Context

It has been observed that most industrialized nations have a set of regulations in place. These regulations are to assure adequate air quality (Giles et al., 2011) that can differ for different sources of pollution (that is industry, transport, heating, etc.).

All these regulations can take the kind of a ban on some agents (Kuhlbusch et al., 2013), like lead in gasoline; (Von Storch et al., 2003) or a technical requirement, like the three-way catalytic converters that are present in cars (Farrauto and Heck, 1999, Merget, and Rosner, 2001).

In addition to it, there are also standards that are set by governmental agencies for levels of pollutants (Bell, Morgenstern, and Harrington, 2011). Clean Air Act in the United States is the most important. (Lawrence, 1971). This regulation was initially passed in the year 1970 and amended periodically ever since (Samet, 2011) that need states to adopt and implement plans in order to achieve and maintain the levels of pollution below federal standards (the United States Environmental Protection Agency, n. d.).

Finally, the selection of a metropolitan area of one or more major sources of energy also plays a very important and large part in the emission of pollutants (Mena-Carrasco et al., 2012).

4.7.2. Demand

It has been observed that the political, economic, and cultural factors can shape the requirements of an urban area in terms of heating, transport, or industry (Lioy and Georgopoulos, 2011). For instance, a city that depends on tourism may give priority to public transportation, green spaces as well as walkability in its center and it will also push the industry to the outskirts.

For example, the cultural values that are linked with car ownership can produce high demand for cars in proportion to the other methods of transportation. Global economic trends such as the increasing dependence on imported goods can also have an impact on the emissions over time (Perez et al., 2009).

4.7.3. Urban Planning

It has been observed that urban sprawl, which in and of itself is reliant on the factors such as topography, demographic shifts, and values can give rise to increased need for cars and longer journeys in cars (Frumkin, 2002).

The demand for private transportation is also affected by design of the public transportation system in a city (Mena-Carrasco et al., 2012). It has been observed that different kinds of public transportation systems will be more or less polluting. This depends on the energy sources of the public transportation system.

4.7.4. Behavior

The behavior of people who are living within an urban area is conditioned by all of the preceding factors. These people may have various options for transportation (such as bus, train, bicycle, motorcycle or car) or for heating. The frequency with which these people use every single option will have impact on the total emissions from every single source (Krizek, 2003, Litman, 2005).

All the above factors are associated both to the source and to emissions. This means that they are determinants for the very existence of the source, or they can also alter the quantity of air pollutants that are emitted. For instance, dense multi-use neighborhoods, by reducing the use of car, work on the emissions. On the other hand, policies banning fuel additives work on the source.

4.8. FACTORS MODIFYING CONCENTRATIONS

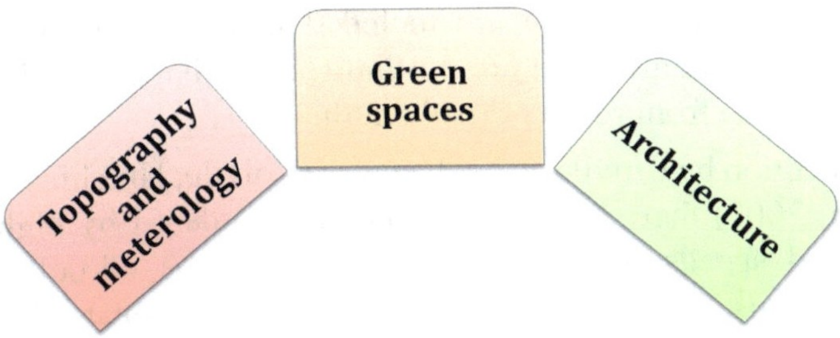

Figure 4.7: Factors modifying concentrations.

The pollutants are dispersed in the atmosphere once the pollutants are emitted. There are various natural and man-made factors that have impact on how pollutants are trapped, dispersed or absorbed.

4.8.1. Topography and Meteorology

In urban areas, the shape of the land can help, or it can also hinder the dispersion of pollutants into the atmosphere, often working in combination with the weather (Wallace, Corr, and Kanaroglou, 2010). It has been observed that valleys lower wind speed, slowing down the rate of dispersion. On the other hand, a coastal location benefits from higher speed of wind (Hertel and Goodsite, n.d.). In a valley, cold weather can make inversion. It can also trap air pollution near to the ground. Heat and sunlight allow the transformation of NO_x and subsequent production of O_3 (Ran et al., 2009).

4.8.2. Green Spaces

Some pollutants such as PAHs and ozone are absorbed by vegetation. In addition to it, it is also capable of trapping and absorbing suspended particulate matter (Vida, 2011). Therefore, the surface area that is devoted to green spaces in a city can have impact on the concentrations of these pollutants.

However, it has been observed that not all green space is equal. For instance, Ouyang, Peng, Chen, Wang, and Jiao (2012) found that the kind of vegetative cover performs a role in how much pollution is absorbed; woodland and tree-shrub-herb settings absorbed more PAH as compared to grassland.

4.8.3. Architecture

There are various factors that are included in the architectural domain. These factors can include (but are not limited to) shape and size, building materials, distance from roads or road width.

It was mention by Colvile and colleagues (Colvile, Hutchinson, Mindell, and Warren, 2001) that narrow roads that are bordered by high buildings mitigate wind speeds, increasing the concentrations of pollutants from vehicle emissions. On the contrary, open roads disseminate the emissions from vehicle in a more effective way.

4.9. FACTORS MODIFYING SPATIAL DISTRIBUTION

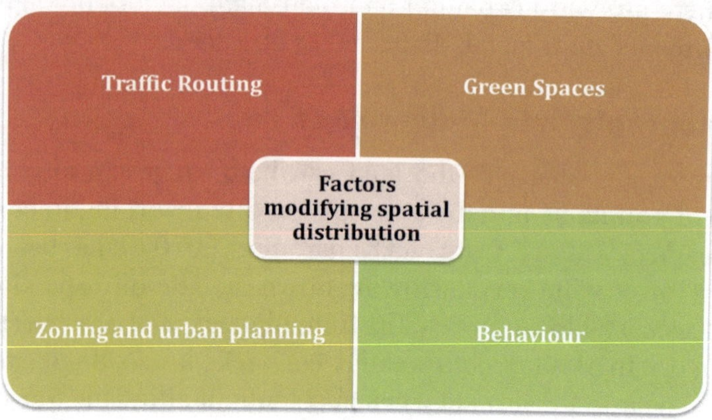

Figure 4.8: Factors modifying spatial distribution.

All the factors that have impact on the emissions, sources, and concentrations have the power to impact the spatial distribution of pollutants (Gilbert et al., 2005, Ragettli et al., 2014a, Ross et al., 2007). This depends on how all these factors act throughout an urban area.

Here, spatial distribution indicates to intra-urban exposure heterogeneity. It does not include the differential distribution that can take place between a city and its surroundings (Ragettli, Tsai et al., 2014). There are few examples that are used to explain what kind of role determinant plays in order to modify the spatial distribution.

4.9.1. Traffic Routing

Figure 4.9: Automobile emits hazardous gases: Impacting air quality.

Source: Image by Wikipedia.

It has been observed that highways and other roads which are having high traffic density may not be distributed evenly throughout the city (Ducret-Stich et al., 2013). The main question of where overpasses and tunnels are built also modifies the distribution of pollution concentrations in the whole city. For instance, some of the vehicles such as trucks may only be authorized to pass through some routes and this, too, will mean that the emissions through vehicles will be higher in certain zones.

4.9.2. Green Spaces

It has to be noted that neighborhoods may have differential surface areas of green spaces (Lakes, Brückner, and Krämer, 2013). They may have differential proximity of green spaces to roadways, or different kinds of land cover in such spaces (De Ridder et al., 2004, Wolch et al., 2014). Any of the aforementioned variations will help in decreasing the concentrations of pollutants to a greater or lesser amount.

4.9.3. Zoning and Urban Planning

Educational institutes such as schools, colleges, and residences, nursing homes may be located far away or close to industrial areas or they can be located on densely trafficked roads. They may be well- or ill-served by the routes of the public transportation.

4.9.4. Behavior

As with behaviors that have impact on the source, behaviors here are extremely constrained by the more upstream factors that are discussed above. It has been observed that different neighborhoods may have access to different methods of transportation or heating sources (Cohen, Boniface, and Watkins, 2014). This gives rise, for instance, to distribution of wood-burning-related pollution. This pollution is centered in neighborhoods without access to gas heating.

4.10. FACTORS MODIFYING PERSONAL EXPOSURE

The exposure of an individual to air pollution is dependent on their behaviors in time and space. It is also called time-activity patterns (Blangiardo et al., 2011, Dons et al., 2011). In addition to it, all the people who are having increased vulnerability to air pollution (like asthmatics) may bring a change in their behaviors to prevent exposing themselves to high levels of air pollution (Buonanno, Stabile, and Morawska, 2014).

4.11. FACTORS MODIFYING VULNERABILITY

There have been many researches that have been conducted on various aspects such as social determinants of health and environmental justice. These researches have demonstrated that the human beings that are most socially and materially deprived are also those who might have to experience a larger impact on their health for the same exposure (O'Neill et al., 2003).

Indeed, two people may be exposed to the same level of air pollution but the health effects for one individual may be more pronounced. This is because of the other social determinants of health like co-morbidities, age, or socioeconomic position.

Vulnerability refers to those people and contextual factors that change the relationship between the exposure of a given person and the health effects that these people may have (Forastiere et al., 2007).

In addition to it, it must also be noted that the health effects that are linked with air pollution can be caused by many other factors, at least some of these factors are addressed by other interventions. For instance, exposure to PM is associated to a short-term increase in heart attacks (Peters, Dockery, Muller, and Mittleman, 2001).

There are various other determinants of heart attacks, which include the level of an individual of physical activity (Ahmed, Blaha, Nasir, Rivera, and Blumenthal, 2012), so a successful physical activity intervention may decrease the vulnerability of a person to suffering an air pollution-related heart attack.

In order to integrate all the vulnerability factors that are documented in the epidemiologic literature, existing frameworks like the progress framework (National Collaborating Centre for Methods and Tools, 2014) can be applied.

This framework recommends a variety of vulnerability factors (for example, education, place of residence, social capital and there are many other factors). These factors can help to examine how several health, as well as non-health interventions, have an impact on the health equity.

Consequently, the responsibility is on the interventions that target the spatial distribution by taking the vulnerability into consideration. Vulnerability will hold an important position in the determination of the spatial distribution of health benefits that are attained by the intervention. Vulnerability helps in attaining a spatially homogenous intervention that offers small health benefits, in a disproportionate manner, to the more vulnerable.

The tool, that is designed, is used to recognize the questions that are associated with the evaluation of the intervention, to identify questions for the improvement of the intervention as mentioned in the table given below.

Table 4.4: Questions to be Aware of for Intervention Evaluation and Improvement, Illustrated for Two Interventions

	General questions	BIXI	LEZ
Sources to pollutants	Question for intervention evaluation: Does the intervention effect the sources of emissions?	Does BIXI decrease the car's use?	Within the geographical area that is defined by the LEZ, has the number of prohibited vehicles reduced? Are there some changes in behavior that are more effective at attaining these impacts than others?
	Question for intervention improvement: Is it possible to change the intervention so as to further decrease the emissions?	What areas or partners should be targeted in order to further decrease the use of car?	How can the most favorable behaviors be encouraged by the intervention?
Pollutants to exposure	Questions for intervention evaluation: Are there some groups that are more exposed to pollutants, because of the intervention? Does the intervention decrease exposure for those groups who earlier had the highest levels of exposure?	Are BIXI users more exposed to air pollutants?	Which groups would be more exposed to pollutants? Is the reduced exposure in the city center offset by a rise in the exposure at the limits of the Zone?
	Question for intervention improvement: What methods or ways can be used so that a decrease in exposure to pollutants are greater for those regions that have higher initial exposure?	Is it possible to construct bike paths that is away from the main roads in order to decrease the exposure of cyclists to roadside pollutants?	Can mechanisms be put in place in those regions that are along the borders of the Zone in order to decrease the impacts of population?
Exposure to health effects	Question for intervention evaluation: Does the intervention have the impact on the populations that are most vulnerable?	It is not pertinent for BIXI	Where are vulnerable zones situated with respect to the main routes? How are they influenced by the LEZ?
	Question for intervention improvement: What can be done in order to decrease the burden for vulnerable zones as well as populations?	It is not pertinent for BIXI	What can be done in order to decrease the burden for vulnerable zones?

Source: Table by Sciencedirect

4.12. CONCLUSION

Air Quality Impact Assessment (AQIA) is the most important aspect in the process of Environmental Impact Assessment (EIA). This is so because Air is the most important aspect of the environment. Since the introduction of the machines in the Industrial Age, the first aspect that was contaminated was Air. It became very important to introduce the AQIA so as to control the level of pollutants in Air. There have been a number of statistical tools that have been devised to imply the policies of AQIA.

A number of methods have been devised by different municipal corporations across the world. These methods differ with various factors that modify and alter the emissions. Several aspects such as Regulatory context, Demand, Urban Planning, and Behavior have a direct impact on the AQIA. With the latest development in the global climate scenario, it is very important to imply the stringent policies that will strictly control the levels of pollutants in the air.

REFERENCES

1. Cartier, Y., Benmarhnia, T., & Brousselle, A., (2015). *Tool for Assessing Health and Equity Impacts of Interventions Modifying Air Quality in Urban Environments*. [online] ScienceDirect. Available at: https://www.sciencedirect.com/science/article/pii/S0149718915000713 (accessed on 4 March 2020).
2. Clark, A., McIntyre, A., Lester, J., & Perry, R., (1984). Air quality impact assessment. *Environmental Monitoring and Assessment*, [online] *4*(3), 205–232. Available at: https://www.ncbi.nlm.nih.gov/pubmed/24257779 (accessed on 4 March 2020).
3. Grobler, N., & Von, R. N., (2018). *Updated Air Quality Impact Assessment for the Proposed Expansion Project at Dundee Precious Metals*. [eBook] Pretoria. Available at: https://s21.q4cdn.com/589145389/files/doc_documents/EN/ESIA/Appendix-F-Air-Quality-Impact-Assessment.pdf (accessed on 4 March 2020).
4. Lustreconsulting.com. (2020). *Air Quality Assessment | Lustre Consulting*. [online] Available at: https://www.lustreconsulting.com/service/air-quality-assessment (accessed on 4 March 2020).
5. Núñez-Alonso, D., Pérez-Arribas, L., Manzoor, S., & Cáceres, J., (2019). Statistical tools for air pollution assessment: Multivariate and spatial analysis studies in the Madrid region. *Journal of Analytical Methods in Chemistry*, [online] 2019, pp. 1–9. Available at: https://www.hindawi.com/journals/jamc/2019/9753927/ (accessed on 4 March 2020).

CHAPTER 5

Water Quality Impact Analysis

CONTENTS

In this chapter, there is a discussion about the strategies of water quality assessment. The objectives of water quality assessment have been discussed in detail in the chapter. There is also a discussion about elements of water quality assessment and levels of water quality assessment in this chapter.

In this chapter, there is a detailed discussion about point and non-point sources (NPSs) of pollution. The examples of point and NPSs of pollution has stated clearly. The different issues in NPSs of pollution has also been discussed in detail.

The impact of wastewater on the surface water quality, and the way decline in surface water quality effects the crops and other different cases has been discussed in this chapter.

Finally, wastewater management and why it is necessary to treat wastewater has been discussed in detail in this chapter.

5.1. INTRODUCTION

In developing countries, the different sources of pollution deteriorate the water quality in rivers. The water resources have become increasingly limited, hard to access, and are often exposed to considerable amounts of wastewater. There are different types of parameters that are used for assessing water portability, among other methods.

The knowledge about these parameters promotes cooperation on a given area of global water quality status and advises the population and administration. The idea of water quality for categorizing water according to its degree of purity or contamination dates back to 1848 in Germany (Dojlido and Best 1993; Horton 1965) to be credited with giving it a first formal definition.

The Water Quality technique has been used to monitor surface and groundwater quality in many countries such as India (Tiwari and Mishra 1985), the United States (Canter, 1996), Indonesia (Gatot and Rina 2011), Canada (CCMC 2001), and Malaysia (Munirah, 2011).

The Similar water quality indexes have been developed in many countries, such as the United States National Sanitation Foundation Water Quality Index (NSFWQI), established by the National Sanitation Foundation (Detroit, Michigan) and the British Columbia Water Quality Index (BCWQI) by the Environment Ministry of Canada (Vancouver, Canada) (Said, 2004). The water quality index is described as a rating that represents the composite influence of various parameters of water quality (Sahu and Sikdar, 2008).

The urban cities in developing countries are growing rapidly and due to demographic and industrial growth in the area over the past decade, the rivers if exist in these cities receive a significant portion of urban, industrial, and agricultural residues, that damages the environment significantly.

5.2. STRATEGIES FOR WATER QUALITY ASSESSMENT

The Assessment of water quality is the overall process of determining the physical, chemical, and biological characteristics of the water, while monitoring of water quality is the compilation of relevant information.

5.2.1. Monitoring, Survey, and Surveillance

Traditionally, the main reason for evaluating the consistency of the aquatic environment has been the need to check whether the water quality found matches expected uses. The use of monitoring has also developed to help assess patterns in water quality and to find out the effects of the release of contaminants in the open, other anthropogenic activities, or waste management operations (impact monitoring) affect the quality. More recently, testing has been carried out to measure nutrient or pollutant fluxes released into lakes and oceans by rivers or ground waters, or across international borders.

The Monitoring of water to assess the aquatic environment's background quality is now also becoming common, as it offers a way of contrast with impact monitoring. It is also used simply to test whether any unexpected changes occur in otherwise uncharted environments, such as the long-range transport of atmospheric pollutants though, the quality of natural water varies greatly depending on the local conditions.

5.2.2. Objectives of Water Quality Assessment

By objectively scrutinizing the actual need for water quality knowledge (that is the need to learn as opposed to it would be nice to know), no assessment system should be initiated. As water resources are typically used for many competing useful applications, the monitoring used to acquire the necessary information will represent the data needs of the various users (Helmer, 1994).

Consequently, according to how many evaluation goals should be met, there are two different kinds of monitoring programs:

- Single objective monitoring which can be developed in order to address only one problem. It requires a basic set of variables such as pH, alkalinity, and some acid rain cations; nutrients and eutrophic chlorophyll pigments, various nitrogen emission compounds, that includes sodium, calcium, chloride, and a few other irrigation compounds.
- Multi-target monitoring that can cover different water uses and provide data for multiple assessment initiatives like the drinking water sources, industrial production, fishing or aquatic life, thereby requiring a wide range of variables.

In drinking water alone, the Commission of the European Communities has a list of over 100 micropollutants. The implementation of the objectives of the evaluation program concentrates on the spatial quality distribution (high number of stations), or pattern (high frequency of sampling) or contaminants (deep inventories). Preliminary surveys are therefore necessary to determine the appropriate focal point of an operating program.

The objective determination method will start with a thorough analysis of all variables and events, directly or indirectly affecting the quality of water.

Inventories shall be prepared with respect to:

- geographical features of the region, including topography, relief, lithology, pedology, climate, land use, hydrogeology, hydrology, etc.;
- use of water, including barriers, canals, water extractions for urban and industrial activities, irrigation, fishing, recreation, fisheries, etc.; and
- Pollution sources (present and expected), including domestic, industrial, and agricultural, as well as their stage of pollution control and waste treatment facilities.

5.2.3. Elements of Water Quality Assessment

There are numerous possible forms of water quality monitoring programs. These should be planned or implemented in compliance with targets set by environmental conditions, water uses (at present and in future), water legislation, etc. Once the goals are set, the monitoring method is decided by an analysis of existing water quality data, often assisted by preliminary surveys.

An important step that is often ignored, if not skipped, is data analysis following the implementation of the different evaluation activities. This should be supplemented by guidelines for water management, regulation of water pollution and finally the improvement or alteration of monitoring activities to local water authorities. There are some basic elements that are applicable to all water quality assessment programs. These are built more extensively, or less, depending on the type of assessment needed.

5.2.3.1. The Basic Elements of an Assessment System

1. Objectives

The hydrological factors water uses, economic development, regulatory policies, etc. should be taken into account. The necessary decisions include whether concentrations or loads should be stressed, or spatial or time distributions, as well as the most suitable monitoring tool.

2. Preliminary Surveys

These are short-term, specific exercises to assess the variation of water quality, the form of monitoring media and pollutants to be considered, and the technical and financial viability of an entire monitoring programs.

3. Monitoring design

It involves choosing pollutant forms, location of the station, frequency of sampling, sampling equipment, etc.

4. Field Monitoring Operations

These include measurements on locations, sampling of appropriate media (water, biota, particulate matter), pretreatment, and storage of samples, identification, and shipping.

5. Hydrological Monitoring

This involves measurements of water intake, water levels, thermal profiles, etc., and should always be linked to the water quality assessment activities.

6. Laboratory Tasks

These include measurements of the concentration, biological determinations, etc.

7. Data Quality Control

The use of scientific quality assurance within each laboratory and among all laboratories involved in the same program and the monitoring of field operations and hydrological data must be carried out.

8. Data Storage Treatment and Reporting

The use of databases, statistical analysis, pattern determinations, multi-factor correlation, etc., and the presentation and distribution of findings in suitable forms that is graphs, tabulated information, data diskettes, etc. are now commonly computerized.

9. Data Interpretation

It includes comparing water quality data between different locations that is water quality descriptors, fluxes, examining patterns in water quality, establishing causal-effect relationships between water quality data and environmental data (geology, hydrology, land use, inventory of polluting sources), and assessing the adequacy of water quality for different uses, etc.

External knowledge may be required for specific problems, and evaluation of the environmental importance of observed changes. The required final stage of assessment activities is the publishing and distribution of data and reports to the relevant authorities, the media and the scientific community.

These decisions should be made at various levels, from local government to international bodies, water authorities and other environmental authorities. Re-designing evaluation processes, enhancing the monitoring program and making it more cost-effective is an important decision. In the evaluation chain the length of each segment is highly variable.

Preliminary surveys and design can take months to years to initiate. A field sampling project may last a few days, shipping may take days, and significant ions, nutrients, and a few other primary variables in a collection of samples typically take a week to chemically analyze.

It may take several months for data to be processed, while analysis and publishing may take from a few months to several years. Nonetheless, fieldwork needs to be carried out for surveillance purposes, data analyses and documentation (including preliminary reports if necessary) within a timescale controlled by operational requirements. In the case of an emergency water quality assessment process, the whole chain should be reduced to a few days, or even less, from concept to understanding.

Evaluation of patterns can often not be given before about ten years, depending on the standard variation and the complexity of the programs.

5.2.4. Levels of Water Quality Assessment

Experience indicates that it is not easy to select the optimum level of assessment required to precisely produce the data needed for specific purposes and goals. The assessment is most often accomplished by tracking practices (that is systematic long-term assessment and observation). The monitoring process will produce the data that is needed for accurate analysis and management decisions but should not lead to a vast collection of unnecessary data that is expensive to obtain but does not contribute to the required understanding of water quality.

In addition, monitoring activities will, by definition, be in line with the country's socio-economic and technical and scientific development. For example, certain variables in water quality (e.g., organic micropollutants, trace elements) need highly trained technicians and expensive laboratory facilities.

The three types can be differentiated in a simplified manner:

- Easy testing based on a limited number of samples, easy analyses or findings, and data analysis that can be performed by the pocket calculator.
- Mid-level monitoring requiring specific laboratory equipment and increased financial resources to increase station numbers, samples, analytical variables, etc. For data processing, Personal Computers (PCs) are recommended.
- For Advanced-level monitoring involving sophisticated techniques and highly trained technicians and engineers. With a growing number of variables per sample, and samples taken, the analytical facilities can perform any pollutant determination needed. Large data storage and handling facilities, such as a mainframe computer, are required for data processing.

All aspects of evaluation will be linked to these three stages, from objective setting to data interpretation. The end goal is to promote monitoring operations from the basic level to the more complicated ones, in line with the increasing complexity of water quality issues and research system performance capabilities.

5.2.5. General Water Quality Monitoring Programs

In theory, there could be as many forms of monitoring systems as there are different goals, water bodies, pollutants, and water use, and any combination of them. It should be mentioned that many countries or water authorities have in the past implemented multi-purpose or multi-objective monitoring programs without carrying out the required preliminary surveys that are deemed necessary. After several years of operation, a thorough review of outcomes has led to the second generation of projects with more specific targets such as impact assessment, trend analysis or organizational management decisions.

The scales (high, medium, low) of all operational characteristics (frequency, density, number of variables, length, lag in interpretation) are given in reference to multi-purpose monitoring, that was taken as a reference point. The significant control features are underscored in bold. Background monitoring (primarily in unpolluted areas) has typically been developed to help understand trend monitoring (long-term time variations) and identify normal, spatial variations.

Models and their associated surveys were generally set up to forecast water quality for management purposes prior to pollution treatment, or to check the effect of a new source of water pollution and are therefore closely linked to operational testing and impact surveys.

In the case of any sudden and unexpected changes in water quality, early warning surveillance is conducted for specific uses, while medium and long-term impact surveys should be accompanied by emergency surveys of a catastrophic event. Many forms of routine testing are often mixed for practical reasons, and some sampling stations will relate to several programs. Multidimensional approaches to determining the water quality have become an unavoidable requirement.

A few water analyses identified the consistency of the aquatic environment around a hundred years ago, but this concept has now reached a level of sophistication that requires multiple factors to be considered simultaneously.

One aspect concerns physico-chemical variables such as nitrates, chromium, and polyaromatic hydrocarbons, another concern guidelines and standards for particular water uses, and another concern various monitoring devices (water, sediments suspended or deposited, colloids, whole organisms, selected biological tissues). The various aquatic ecosystems (streams, rivers, wetlands, lakes, reservoirs, all forms of groundwater) react differently to contamination with different variability scales in terms of time (from minutes to years) and space (vertical, transverse, longitudinal, upstream to downstream).

Additionally, it is necessary to consider aquatic organisms, from bacteria to mammals, including various types of biological characteristics, such as the structure of community or bioassays. In fact, in all its dimensions, water quality is never considered. In most situations, the approach to the assessment is decided by the perceived importance of the aquatic environment, the goals of the project, and the available human and financial resources.

For example, lake eutrophication is considered mostly through nutrient analysis, biomass estimates using chlorophyll measurements, dissolved oxygen profiles, and a few phytoplankton indicator species. Monitoring practices are also highly dependent on the level of aquatic habitat deterioration.

Background testing establishes the natural concentrations and verifies that there were no xenobiotic organic substances at this point. To determine the subtle changes that characterize anthropogenic impacts, careful monitoring of early change is required. The basic issue at this point is what issues should be considered within the context of the basic monitoring that is regularly carried out on all water bodies.

Water quality problems are often studied through explanatory or predictive models, which involve extensive, detailed monitoring and are required to determine the basis for mitigating measures and regulations. When the issue is more fully understood, the issue can be based on environmental impact evaluations before control measures are effective. If they aren't effective, loss of water quality becomes a major concern and early-warning monitoring may need to be implemented. Such a method of monitoring is often used at drinking water intakes to warn about upstream incidents of accidental contamination.

5.3. POINT AND NON-POINT SOURCES OF POLLUTION

5.3.1. Categories of Pollution: Point Source

The U.S. Environmental Protection Agency (EPA) describes point source pollution as any single recognizable source of pollution from which contaminants are emitted, such as a smokestack pipe, dyke, ship or factory (Hill, 1997).

The two common types of point sources are the factories and sewage treatment plants (STPs). The factories, along with oil refineries, pulp, and paper mills, and manufacturers of chemicals, electronics, and automobiles generally discharge one or more pollutants into their ejected waters that are known as effluents. Some factories directly discharge their effluents into a waterbody.

Some process it themselves before it is released and yet others send their waste for treatment to STPs. The STPs process human waste and transfer the treated effluent to a river or stream.

Another way that certain factories and STPs manage waste material is by combining it in a combined sewer system with the urban runoff. Runoff corresponds to stormwater which flows through surfaces such as driveways and lawns. As the water hits certain surfaces, contaminants, and pollutants are collected. This untreated, contaminated water then runs straight into a sewer system.

The combined sewer system might not be capable of handling the volume of water when it rains heavily, and some of the combined wastewater and raw sewage may leak from the system, discharging directly into the nearest water source without treatment. This combined sewer overflow (CSO) is considered a source of point pollution and can cause serious damage to human health and the environment.

The unregulated discharges from point sources might cause water contamination and unsafe drinking water and could restrict activities such as fishing and swimming. Some of the chemicals released by point sources are harmless but others are toxic to humans and wildlife. Whether such a discharged chemical is hazardous to the aquatic environment depends on a number of factors, that include the type of chemicals, their composition, timing of their release, weather conditions and the species residing in the area.

Other sources of point source pollution are large farms which raise livestock, such as cows, pigs, and chickens. Such farm types are known as concentrated feeding operations (CAFOs) for animals. If they do not handle waste materials from their animals, these contaminants can then enter nearby water sources as raw sewage, contributing significantly to pollution levels and its rates.

To regulate point source discharges, the National Pollutant Discharge Elimination System (NPDES) was established by the Clean Water Act in the United States of America. Under the NPDES scheme, manufacturers, STPs, and other point sources must receive approval of the state and EPA before they can discharge their waste or effluent into any water body.

Before the discharge, the point source must use the latest available technology to handle the effluents and rising levels of pollutant.

A second, more stringent group of controls can be imposed on a point source to secure a specific water body if necessary. Point source water pollution includes pollutants which are emitted from any known, specific source, such as a pipe, ditch, drain, channel, tube, duct, well, container, or vessel.

The Non-point source (NPS) water pollution, on the other hand, is caused by widely dispersed and disconnected contamination sources such as rain and snowmelt runoff, spills, leaks, and sediment erosion. The Regulated point source contaminants include waste, dirt, rocks, chemicals, bacteria, suspended solids, heavy metals, pesticides, and more.

Examples of point sources include STPs; oil refineries; paper and pulp mills; producers of chemicals, vehicles, and electronics; and factories. Regulated point-source contaminants include waste, dirt, rocks, chemicals, bacteria, suspended solids, heavy metals, pesticides, etc.

Figure 5.1: Industrial plant discharge into the river.

Sources: Image by Wikimedia.

5.3.2. Categories of Pollution: Nonpoint Source

The majority of the nonpoint source emissions are caused by runoff. The water absorbs and assimilates any contaminants it comes into contact with when rain or melting snow flows over and through the ground (USEPA, 2004). For example, after a heavy rainstorm, water flows through a parking lot and absorbs oil left by cars driving and parking on the asphalt. They are actually looking at nonpoint source contamination when they see a rainbow-colored sheen on water flowing across the surface of a road or parking lot.

The runoff then flows across the bottom of the parking lot, and it will most likely eventually drain into a lake. The water flows into a larger stream downstream, and then into a lake, river, or ocean. The contaminants can be quite dangerous in this runoff, and several of their sources. Usually one can't point to one discreet nonpoint source pollution location as one can with a factory discharge pipe.

Nonpoint source pollution affects not only wildlife but also the economy. The United States Coastal and marine waters sustain 28.3 million jobs, produce 54 billion dollars in goods and services from activities such as shipping, boating, and tourism, and add 30 billion dollars in recreational fishing alone to the US economy (Lee worthy, 2000). When pollution contributes to massive fish die-offs and dirty-looking water, this region and others like it will suffer severe financial losses.

Pollution from nonpoint sources threatens the beauty and wellbeing of coastal lands and waters. If these areas' physical and environmental well-being are reduced, people will, of course, find it less desirable to visit the coast. Beaches won't offer the relaxation and leisure activities that many people expect to experience.

One can see the way nonpoint source pollution plays an indirect, but powerful role in the contributions of visitors to the economic status of a coastal community the population in many coastal communities is also growing at a rapid pace, and the value of waterfront property also depends on environmental and ecological conditions.

The more than tolerable nonpoint source emissions affect the overall quality of life and can then drive down property values. If nonpoint source pollution continues to emit in coastal communities surrounding waters, their economies and social conditions can deteriorate rapidly.

Although the concentration of some contaminants from runoff may be smaller than a point source concentration, the total amount of a pollutant distributed from nonpoint sources may be higher because the pollutants come from many locations.

With increased control over point source pollution, scientists have started to focus on nonpoint source pollution, the manner in which it affects the environmental quality and, more importantly, the way it can be controlled. Nonpoint source pollution is hard to control because the emissions come from multiple locations. It also varies with time regarding the flow and the pollutant forms this produces.

Pollution from nonpoint source water leads to the degradation or depletion of water bodies. Nonpoint source pollution is caused by diffused and disconnected emission sources. Such causes may include runoff from rain and snowmelt, spills, leaks, river pollution, pesticide, and pesticidal residues, sewage overflows, croplands, failed septic tanks, and more. Conversely, water pollution from a point source is caused by specific and recognizable sources, such as drains, ditches, mines, sewage plants, and more.

These are included in the nonpoint source of pollution:

- Animal production and feedlots;
- Farming and cropland;

- Shoreline erosion;
- Land development;
- Sewage overflow;
- Solid waste disposal;
- Coal mining;
- Mineral extraction;
- Atmospheric deposition such as dropping particles; and
- Runoff containing chemicals, sediments, fertilizers, and phosphorus.

Figure 5.2: Non-point source pollution.

Sources: Image By Flickr.

5.3.3. Issues Associated in Non-Point Source Pollution

The deterioration of land and water resources and their spatial and temporal variations exacerbate the problem of a balance between the resisting forces responsible for maintaining a farmland's equilibrium. Following a few decades of continuous soil erosion, crop rooting depth may become minimal, resulting in lower productivity. In addition to erosion, certain farming practices may also result in a reduction in land productivity.

Where the efficient use of fertilizers is not practiced, a decrease in nutrient content and decalcification can result in chemical degradation.

1. Stormwater Runoff

The stormwater runoff is a major component of non-point source pollution (NPS), the largest remaining source of pollutants in the water of our Country, according to the United States Environmental Protection Agency (USEPA). Pollutants in storm-water runoff emerge from multiple, dispersed sources and are spread by runoff from a whole drainage region or watershed, they can be difficult to identify, control, and handle. Stormwater is the rain, snow, and sleet water that flows into the storm drain down our gutters.

Stormwater starts off clean, stormwater flows into our rivers, reservoirs, and streams directly. It is basically never treated. So, everything that is collected by stormwater from the surface of the ground, highways, sidewalks, parking lots, building sites, business parks, etc. goes straight into the drain.

It is transported via gutters, storm drains, canals, and means of drainage and finally ends up unchecked in our local rivers and streams. It is estimated that more than half of the contamination in the rivers of our Nation comes from the runoff of stormwater.

The best way to improve the quality of stormwater is to manage the source first, don't let runoff get contaminated. The contamination of surface water and groundwater caused by stormwater runoff will affect both the quantity and quality of drinking water sources needed to sustain existing communities and industries, and the recreational use of streams, rivers, and lakes.

2. Land Degradation

Soil erosion is the major cause of land source degradation. The process involves the land detachment and soil transport. According to information published by the Ministry of Agriculture in 1980, there are severe environmental degradation in India as many as 175 million hectares (constituting 53% of the geographical area of India). About 60% of the cultivable area requires steps for soil conservation.

So, the single biggest threat, in fact, is the increasing mismanagement of our soil and land resources in order to achieve our agricultural dreams. About 150 million hectares were trapped by runoff and wind in the vicious circle of erosion, as much as more than 90 million hectares are impacted by water-related erosion. One recent estimate is at 12,000 million tons per year places the loss of topsoil by water action.

It is worth noting that building up 1 cm of the topsoil takes 200–400 years for nature. However, 25 million hectares of land has been eroded due to ravines and gullies, crop changes, salinity or alkalinity, waterlogging, etc. According to the Agriculture Department, every year the country loses 30–50 million tons of food grains due to the loss of fertile topsoil of about 85 million hectares of degraded agricultural land. The problem is serious enough for a country that supports 15% of the world's population in 2% of the world's land area. (Substitute al., 1995)

3. Sedimentation

In addition to the loss of productivity, the accumulation of the detached soil has become a major cause of recurrent reservoir sedimentation. Sediment-shaped degraded soil is borne along with the storm runoff of water storage bodies, including ponds, lakes, dams, etc. The sediment gets stored in these types of reservoirs over time, leading to a decrease in their storage capacity and usable life.

The recent observations have revealed the alarming fact that the sedimentation of the reservoir resulting from the degradation of the watersheds is on a multiple increases compared to the rate assumed at the time of the design of the projects. This leads to depletion of the wetlands, which leaves fertile fields sterile, decreases dam storage capacity and therefore reduces their service life.

4. Nutrients

Non-point nutrient sources are often derived from agricultural activities (EPA, 2004). The Excess nutrients that are added to crops in the form of fertilizers are washed away in runoff, usually during rainfall. Even nutrients come from urban and suburban areas, from sources like lawn fertilizers and even pet waste. Nitrogen and phosphorous both derive from sources from the atmosphere. Controlling the inputs of nutrients proves to be very difficult because the nutrients often come from multiple sources which are hard to recognize and monitor.

Table 5.1: Types of Pollutant and Its Effect

Types of pollutants	Where it came from	What it can do
Sediment (dirt)	Construction, agriculture, and mining sites not correctly using BMPs	Cover fish and aquatic organisms' habitat, destroy oxygen enriching riffle areas, shield other pollutants from disinfection from sunlight
Organic Matter	Yard clippings, failing septic tanks, not picking up after your pet	When organic matter starts to decompose, it uses oxygen. If Enough matter starts decomposing it can use up all of the oxygen and can use up all of oxygen and kill of the aquatic organisms. Also, animal and human waste can contain diseases and can pose as a human health risk
Nutrients (Phosphorus and Nitrogen)	Over-fertilizing lawns, soaps, animals, and human waste	Create algal blooms that can destroy ecosystems. When algae start dying, it uses up the oxygen and kills of the aquatic organisms.
Fuel, oils, and other automotive fluids	Outdoor car washing, Placing motor oil into storm drain, parking lots	Create public health hazards, block oxygen from diffusing into the water, halt photosynthesis, Destroy habitat, clog fish gills and smother other organisms
Poisons	Improper use of pesticides and herbicides	Kill aquatic organisms, Kill aquatic plants. Plants are needed for aquatic habitat and food
Inanimate objects	Litter, illegal dumping of tire and trash	Create Flooding problems, cause erosion of stream bank, destroy habitats.
Others, Such as Metals and Industrial waste	Mostly illegal dumping, but can come from land-disturbing activities	These items indirectly cause health problems by creating fish consumption limits and hinder drinking water supplies, if not properly dealt with.

Sources: Table from Shodhganga.

5.3.4. Causes of Non-Point Source Pollution

1. Pesticides

Pesticides that are applied to farm fields and roadsides and to the lawns of homeowner's spill into local streams and rivers, or flow into groundwater, contaminating the freshwater that fish swim in and the water that the humans drink. It is right to think this is mostly a farming issue, but homeowners apply even more chemicals to their lawns on a square foot basis than farmers do to their fields. But agriculture is a major contributor to this issue.

2. Fertilizers or Nutrient Pollution

Most causes of pollution, that includes waste, manure, and chemical fertilizers, comprise such nutrients like the nitrates and phosphates. Nutrient-type water contamination is also caused by the accumulation of atmospheric nitrogen that is from nitrogen oxides. Nutrients over-stimulate the growth of aquatic plants and algae in excess levels.

The excessive development of these types of species obstructs our rivers and blocks light into deeper waters while the organisms are alive; when the organisms die, they use dissolved oxygen as they decompose, creating oxygen-poor waters that only sustain reduced marine life. Those regions are often referred to as dead zones. Nutrient contamination in estuaries and deltas is a particular problem, where the runoff that was aggregated by watersheds is eventually deposited at the mouths of major rivers.

3. Oil, Gasoline, and Additives

Shipping is one of these non-spill sources of oil contamination in water: oily waste disposal and oil-contaminated ballast water and wash water are all major sources of marine emissions, and ship and boat engine drips add their share. Oil and gas exploration and mining activities can also pollute coastal waters and groundwater. As for additives to oil and coal, leaking storage tanks is a major problem. The U. S. Environmental Protection Agency (EPA) reports that nearly 100,000 storage facilities leak pollutants into groundwater.

4. Sediment

The root systems that previously held soil in place die when trees are cut out, and sediments are free to run off into surrounding streams, rivers, and lakes. Therefore, not only does clear-cutting have severe effects on plant and animal habitats in the forest, but the increased amount of sediment flowing off the land into nearby water sources has a serious impact on fish and other aquatic life. Poor farming practices that leave soil exposed to the elements also contribute to water contamination of the sediments.

5. *Sewage*

Approximate 90% of wastewater is dumped directly into rivers and streams in developing countries without treatment. Untreated sewage, poorly treated sewage or overflowing from under-capacity sewage treatment facilities can carry disease-bearing water into rivers and oceans, even in modern countries. Leakage of septic tanks and other sewage sources can cause pollution of the groundwater and streams.

Figure 5.3: Runoff of soil and fertilizer.

Sources: Image by Wikimedia.

5.4. IMPACT OF WASTEWATER ON SURFACE WATER QUALITY

One of the major challenges that is being faced by the world, is the availability of the freshwater and about one-third of the world's drinking water demand is derived from surface sources such as rivers, dams, reservoirs, and canals.

Such types of water sources also act as the best sinks for the disposal of household and industrial waste. The developing countries major threat to sustainable water supply is contamination of available water supplies by pollution.

There are majority of developing countries communities that still rely on untreated or inadequately treated surface water, such as rivers and lakes, for their daily supply. They have no or restricted access to decent sanitation facilities and pose a high risk for waterborne diseases. The incidents of waterborne diseases in developing countries are increased day since the start of this decade.

Humans have used the surface water for several purposes. It serves as a source of after-treatment drinking water and as a source of untreated domestic water, especially in developing world rural areas. It has been used by farmers for irrigation purposes, and fishermen get their jobs from harvesting fish in so many sources of freshwater. It is used for swimming and is also used as tourist attraction centers.

Therefore, surface water should be protected from contamination. Pure and partially treated wastewater are significant point sources of freshwater contamination. The domestic and industrial wastewater release has resulted in increased freshwater contamination and clean water supply depletion.

Most quantities of wastewater produced in developing countries are not treated in any way. Various forms of wastewater treatment facilities (WWTFs) are present in a few urban centers, but most of them contain untreated effluents that are disposed of on freshwater courses. The suitable availability of potable water and improved sanitation services have been reached in several of the developed countries of the world.

The Strict environmental regulations and enforcement control discourage excessive pollution of freshwater supplies. Good waste management technologies and increased awareness of environmental protection have made an immense contribution to the success story. This has contributed to fewer reported cases of waterborne diseases as compared with developing countries.

Most citizens in the developing countries of the world already depend on untreated surface water as their primary source of home water supply. This is because there is either an incessant flow of potable water or inadequate water supply systems. Across rural areas this issue is complicated.

Due to population growth and accelerated industrialization surface water is coming increasingly under undue stress. The easiness of surface water accessibility makes them the best choice for discharging wastewater. Wastewater, consisting of several micro-organisms, heavy metals, nutrients, radionuclides, pharmaceutical, and personal care products, all find their way to surface water resources that cause irreversible damage to the aquatic ecosystem and humans as the esthetic value of such water is compromised.

Such contaminants reduce the supply of usable water, raise the purification costs, contaminate the aquatic resources and impact the food supplies. Pollution combined with human water demand impacts the biological diversity, the functioning of the habitats and the natural resources of aquatic environments on which society depends.

In most developing countries, urban areas have several wastewater management systems, some of which are very effective and meet international standards, but many others are plagued by poor designs, maintenance problems, and expansion, that comprises inadequate investment in wastewater management.

Many rural and poor areas often have no wastewater management systems at all. The effluents from large and small industries are generally channeled to surface water channels, which often contribute to pollution loss of biodiversity in the aquatic ecosystem, and probably the risk to human health. The most commonly used strategies for regulating and reducing environmental pollution are environmental quality and anti-pollution legislation.

In the majority of countries, the government has passed environmental laws and implemented them through its administrative structures. The pollution is also limited by the use of criminal sanctions, but enforcement of these environmental laws remains inadequate.

The enforcement of environmental laws in developing countries, as in all the developing countries, is suffering major setbacks because of inadequate technical experts, inadequate funding, corruption, and low sanctions deterrent effects.

5.4.1. Surface Water Quality

Surface water is one of the most affected habitats on Earth and its changes have resulted in significant ecological degradation, such as a decrease in water quality and availability, severe flooding, biodiversity extinction, and adjustments in the distribution and structure of aquatic biota, rendering surface watercourses unsustainable in the supply of products and services.

For example, a health of the river system is affected by a variety of factors that comprise of geomorphology and geological formations, physicochemical, and microbial water quality, hydrological regimes, and in stream and riparian habitat nature. The water quality is characterized by water's chemical, physical, and biological characteristics, which determine its fitness for a variety of uses and for the protection of aquatic ecosystem health and integrity.

Every aquatic ecosystem has a natural inclination to adapt and offset changes in water quality parameters by diluting and biodegrading certain organic compounds. However, when this normal buffering ability of the aquatic ecosystem is reached on a continuous basis due to the introduction of different classes of pollutants from point and nonpoint sources, water pollution sets in. In developing countries, like most other developing countries in the world, surface water is usually used primarily in rural areas for domestic, recreational, and agricultural purposes.

The water quality is affected by both natural and anthropogenic processes. The natural water quality usually varies from place to place, depending on the seasonal changes, climatic changes, and the types of soils, rocks, and surfaces it passes through it. The different type of human activities such as farming, urban, and industrial growth, mining, and recreation significantly alter the nature of natural waters and shift the capacity for water use.

Declining water quality will result in higher treatment costs for drinking and industrial process water. The use of poor-quality water for farming activities can affect crop yield and cause food insecurity. Around the world, the presence, transportation, and fate of heavy metals and organic compounds that are harmful and widespread in water bodies is a cause for grave concern.

The release of chemicals contained in wastewater can pollute the groundwater. The wetlands and riverbeds are threatened by increased impoundment of sediments and the presence of harmful and persistent chemicals. These pollutions can persist long after the cessation of their original sources.

The presence of toxic substances can have a detrimental effect on the health of the aquatic ecosystem. This is further compounded in the water with a high pathogen population. The use for domestic and other purposes of microbiologically contaminated water is detrimental to human health and society at large.

Such type of conditions may also have an effect on wildlife that uses surface water to drink or as habitat. In general, physical (turbidity, electrical conductivity, temperature, total dissolved solids, color, and taste), chemical (pH, COD, BOD, non-metals, metals, and persistent organic pollutants, POPs) and biological (fecal coliform, total coliform, and enterococci count) analyzes are performed for measuring water quality.

5.5. WASTEWATER MANAGEMENT

Wastewater is water that has modified its physical, chemical or biological properties as a consequence of the application of certain chemicals that make it unsafe for certain uses such as drinking. The daily life activities of human beings are mostly water-dependent, and thus discharge waste into water.

Some of the pollutants include solid waste and urine, hair shampoo, blood, food scraps, fat, laundry powder, toilet paper, pesticides, detergents, household cleaners, garbage, micro-organisms (germs) that can cause illness and harm to the ecosystem.

Much of the water supplied is known to end up as wastewater which makes its treatment very necessary. Wastewater treatment is the process and technology that is used to remove most of the pollutants present in wastewater to ensure a safe atmosphere and good public health. Accordingly, wastewater management means managing wastewater to protect the environment and ensure public health, economic, social, and political soundness (Metcalf and Eddy, 1991).

5.5.1. Objectives of Wastewater Treatment

Wastewater treatment is very necessary for the wellbeing of the environment.

It is more vital for the:

1. Reduction of Biodegradable Organic Substances in the Environment

The organic substances such as carbon, nitrogen, phosphorus, Sulfur in organic matter needs to be broken down by oxidation into gases which is either released or remains in solution.

2. Reduction of Nutrient Concentration in the Environment

The nutrients such as nitrogen and phosphorous from wastewater in the environment enrich water bodies or render it eutrophic leading to the growth of algae and other aquatic plants. These plants deplete oxygen in water bodies, and this hampers aquatic life.

3. Elimination of Pathogens

The organisms that cause disease in plants, animals, and humans are called pathogens. They are also known as micro-organisms because they are very small to be seen with the naked eye. Examples of micro-organisms include bacteria (e.g., *Vibro cholerae*), viruses (e.g., enterovirus, hepatitis A & E virus), fungi (e.g., *Candida albicans*), protozoa (e.g., *Entamoeba hystolitica*, *Giardia lamblia*) and helminthes (e.g., *Schistosoma mansoni*, *Asaris lumbricoides*).

These micro-organisms are excreted in large quantities in faeces of infected animals and humans (Awuah and Amankwaa-Kuffuor, 2002).

4. Recycling and Reuse of Water

Water is a scarce and finite resource which is often taken for granted. In the last half of the 20th century, population has increased resulting in pressure on the already scarce water resources. Urbanization has also changed the agrarian nature of many areas.

The Population increase means more food has to be cultivated for the growing population and agriculture as we know is by far the largest user of available water which means that economic growth is placing new demands on available water supplies. The temporal and spatial distribution of water is also a major challenge with groundwater resources being overdrawn. It is for these reasons that recycling, and reuse is crucial for sustainability.

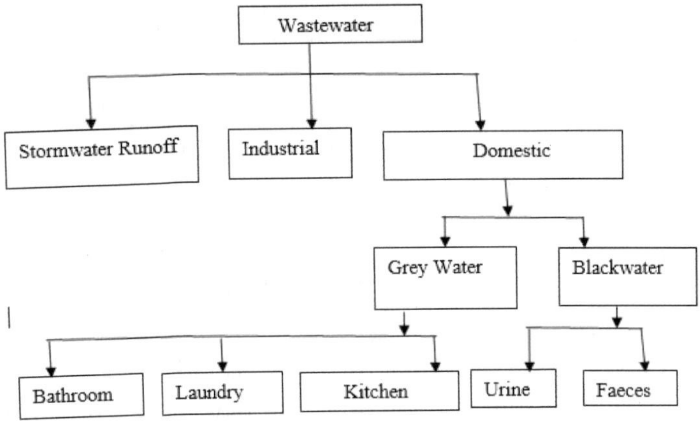

Figure 5.4: Types of wastewater.

5.5.2. Wastewater Treatment

Wastewater contains all water used in homes and businesses, including stormwater and land runoff, which must be handled before it is released into the environment to prevent any damage or danger to the environment and human health. The main purpose of wastewater treatment is to protect human health and prevent degradation of the environment through the safe disposal of the domestic and industrial wastewater generated during water use. One of the aims of wastewater treatment is to recycle wastewater for irrigation reuse, thus maintaining water resources which are scarce in the world's arid and semi-arid regions.

In ancient times, the wastewater was given no specific treatment. Then, wastewater was channeled into gutters and canals from buildings into waterways, which eventually ended up in rivers, streams, lakes, and oceans used by humans. This natural dilution-based treatment process was likely sufficient due to a smaller population and low population density as well as human activities, resulting in lower pollution load compared to the current times.

The population growth and industrial growth led to the production of a high quantity of untreated wastewater channeled as raw water into water bodies. Eutrophication, fish killing, and outbreaks of cholera is commonly reported in communities using contaminated water for domestic and other purposes. It necessitates consideration of a more modern wastewater treatment system.

The treatment facilities for wastewater were initially designed to remove or decrease conventional pollution parameters (BOD; COD, total suspended solids, and nutrients) from the wastewater stream so that the final effluents do not constitute new pollution sources. It has been discovered, however, that the organic wastewater charge contains high levels of a number of hazardous organic contaminants and therefore additional treatment steps and control measures are becoming very necessary. The drainage quality varies depending on the types of influences provided by WWTFs such as domestic drainage, dry, and wet atmospheric deposition, urban runoff with traffic-related emissions, or agricultural runoff.

As industrial wastewater is included in the raw water stream that enters a wastewater treatment facility, the range of contaminants becomes broader. The wastewater effluents have recently been shown to contain emerging organic contaminants, such as persistent organic pollutants (POPs), brominated flame retardants, per-fluorinated compounds, and pharmaceuticals that are not extracted during the treatment process.

Wastewater treatment technology is changing rapidly to meet the challenge of the present day. Urbanization is increasing at an unprecedented rate in many countries, and such growth is often unbalanced with much of the disposable municipal spending committed to high-profile infrastructure with waste disposal and management dropping well into the list of funding allocation priorities.

A study conducted by Save Water in Developed Countries found that less than half of the wastewater treatment plants (WWTPs) in the developing countries handle wastewater that they receive at a healthy and appropriate level. The Wastewater health risk usually comes from microbial pathogens, nutrient loads, heavy metals and certain organic chemicals.

The bacteria are the most common pathogens commonly found in treated wastewater and cause a number of infections and diseases, particularly among young, pregnant, immune-compromised, and elderly people.

Many wastewater treatments plants in developing countries dispose of their effluents directly into nearby rivers or streams, which the surrounding villages use for their different water needs. From their research on wastewater treatment plants in the Limpopo Province of South Africa, Ogola et al.

demonstrated that wastewater is rarely treated to acceptable standards, and this was further verified by Edokpayi and Pindihama et al.

Their findings suggest that insufficient investment in wastewater treatment infrastructure, lack of skilled workforce, poor planning or corruption could have led to the poor performance of WWTFs in that area. The Wastewater needs to be properly prepared before disposal or reuse to protect the water bodies receiving from contamination. The dumping of poorly treated wastewater usually affects downstream water users and contaminates groundwater.

The Waste stabilization ponds (WSPs) are commonly used to provide the smaller towns and communities with an efficient and low-cost means of handling domestic wastewater. Using WSPs is beneficial over traditional WWTPs, as they are very easy to build, run, and maintain and do not necessarily require professional manpower. Jagals et al. however, conducted a study on WSPs in the Free State Province of South Africa and their results revealed that most municipal WSPs performed less than the standard required.

They suggested that a regular monitoring system be introduced. Wastewater has also been identified as a potential source of heavy metals, polycyclic aromatic hydrocarbons (PAHs), and soil, surface water, wastewater, and groundwater contamination. In most WWTPs, the insufficient storage facility also gives room for untreated wastewater losses into surrounding lands and rivers, especially during heavy rains and flooding.

Weber et al. demonstrated from their studies the presence in the wastewater of a wide range of contaminants from emerging concerns even after the conventional treatment process. Pesticides, PAH, and pharmaceutical and personal care products are among those pollutants. Bundschuh et al. reported on the impacts of wastewater, that the presence of micro-organisms was largely associated with concerns, while other toxic and persistent components such as heavy metals and POPs were not adequately considered.

The treated wastewater also contains certain contaminants that are not eliminated in treatment systems, such as POPs and heavy metals. The use of such water for irrigation will result in the accumulation of those pollutants in soils that can be bioavailable for plant and animal uptake. While providing man with a route through the food chain.

After long-term applications, they have also biological effects on soil fauna and flora. The growing of vegetables in wastewater irrigated soils containing high concentrations of toxic metals typically absorbs these metals and accumulates them in edible and non-edible parts of the vegetables in amounts large enough to pose potential health risks to both the animals and humans consuming these metal-rich plants. The heavy metals have particular characteristics which make them toxic even at very low concentrations.

Through various environmental media, they are nonbiodegradable and permanent, and can accumulate through plants and animals. The oral route was established as the principal mechanism through which heavy metals reach the human body. Consumption of food crops from wastewater irrigated farmlands and ill-treated wastewater effluents could make people who feed on them at risk for several diseases, some of which only become evident after many years of exposure.

Wang et al. demonstrated the accumulation of PAH in wastewater irrigated soils, such that the concentration of PAHs was found to be higher in soils very close to the main entrance to the wastewater and gradually decreased at a distance from the plant. Once these pollutants are released into the environment they can persist for a long time to come. PAHs can have external and internal effects on both humans and animals.

They cause many inflammations on the skin which are often related to itching and irritation. These are also cancer-causing and endocrine disruptors. In developing countries, most research on the effect of wastewater on the receiving environment are often limited to the microbiological composition of the discharged effluent.

Figure 5.5: Wastewater treatment plant.

Sources: Image from Pixabay.

5.6. CONCLUSION

The countrics are urbanizing at rapid pace and these cities are backbone of the economies of many developing countries. The large numbers of people are migrating to urban cities and are settling there. They all need clean and safe water for drinking and other purposes. The Scarcity of water and water pollution constitute a big challenge for water managing bodies in the developing countries. In developing countries, the water quality standards are low in comparison to developed countries.

The sewage water can be treated by wastewater plants to reduce the impact on the environment and human health. The surface water of rivers should be clean for the wellbeing of aquatic life.

REFERENCES

1. Amoatey, P., & Bani, R., (2011). *Wastewater Management*. [online] Research Gate. Available at: https://www.researchgate.net/publication/221911472_Wastewater_Management (accessed on 4 March 2020).
2. Ballotpedia, (n.d.). *Point Source of Water Pollution-Ballotpedia*. [online] Available at: https://ballotpedia.org/Point_source_of_water_pollution (accessed on 4 March 2020).
3. Chapter 6: Non Point Source Pollution. (n.d.). [eBook] p.2. Available at: https://www.who.int/water_sanitation_health/resourcesquality/wqachapter2.pdf (accessed on 4 March 2020).
4. Edokpayi, J., Odiyo, J. and Durowoju, O., 2017. Impact of Wastewater on Surface Water Quality in Developing Countries: A Case Study of South Africa. *Water Quality*, [online] p.2. Available at: <https://www.intechopen.com/books/water-quality/impact-of-wastewater-on-surface-water-quality-in-developing-countries-a-case-study-of-south-africa> [Accessed 4 March 2020].
5. Guettaf, M., Maoui, A., & Ihdene, Z., (2014). Assessment of water quality: A case study of the Seybouse River (North East of Algeria). *Applied Water Science, 7*(1), 295–307.
6. Meybeck, M., Kimstach, V., & Helmer, R., (1996). *Chapter 2-Strategies for Water Quality Assessment*. (2nd edn., p. 2). [eBook]. Available at: https://www.who.int/water_sanitation_health/resourcesquality/wqachapter2.pdf? (accessed on 4 March 2020).
7. Oceanservice.noaa.gov. (2020). *NOAA's National Ocean Service Education: Nonpoint Source Pollution*. [online] Available at: https://oceanservice.noaa.gov/education/kits/pollution/04nonpointsource.html (accessed on 4 March 2020).

CHAPTER 6

Impact on Soils and Land Environment

CONTENTS

There is a growing concern that the environmental impact on the soil and land would adversely affect the agricultural production. Soil is the fine earth particles that lie on the surface of the earth. Whereas, the land is the solid part of the earth's surface on which the life grows. This chapter gives insights about the environmental impacts on the soil and land environment. In the mid of the chapter, soil erosion, its causes, functions of soil and major soil processes are discussed. Going forward, the environmental impact on the land, its relationship with the humans, concepts of land are discussed. At the end, the chapter explains the difference between land cover and land use and the global issue of land degradation is discussed. Additionally, it has explained the land degradation control measures at global level.

6.1. INTRODUCTION

Soil is basically the fine earth that covers the land surfaces. The soil cover is the result of the weathering of rock materials and the accumulation of mineral matter carried away by water, wind, or ice. The weathered mineral material is an added organic material which forms a distinctive feature of soil. The organic material can be living or can be dead.

The dead organic matter includes the freshly added dead plant roots, leaf, and other plant litter, dead fauna, and organic material. There are various stages of decomposition from little modified relatively fresh materials to the complex decomposed material called humus. The distinctive characteristic of soil is achieved by this mixture of mineral and organic material.

Throughout the surface of the earth, there is varying types of soil that is a combination of mineral and organic matter. The responses of the different types of soil have different responses to the environmental conditions. The balanced system is maintained by the soil, its constituents, the plant life it supports, the rock on which the soil lies, and the climate it experiences.

There are different meanings of soil depending upon different context. According to a simple definition of soil, Soil is the material that plants grow in and which provides them with physical support, water, and nutrients.

The soil for the engineers is the finely divided and relatively loose "rock" material at the earth's surface, and this overburden is often considered an inconvenience because it must be removed. The geologist calls the layer of the soil the regolith and geological investigations frequently begin below it.

As per hydrologists, the soil is a large reservoir storing water to supply streams and rivers. However, there is a common definition of the soil, and as which, the soil is the medium for plant growth and the provision of food and fiber directly or through the intermediate stage of animals.

Soil is important for the plant growth and for the support animal and human activity. Throughout the growth of the plant, soil acts as a reservoir for nutrients and water providing the plants' needs. There may be an environment that results in the breakdown and immobilization of materials added to the surface such as fertilizers and pesticides and waste products such as sewage sludge, animal wastes and slurries, and composted refuse materials.

The interactions of the biological, chemical, and physical environments result in the transformation of materials. The soil is a complex dynamic system and it renders the harmful materials less dangerous and immobilizing others as a result of the interactions between these added materials and the organic and inorganic soil constituents.

The dangerous material is breakdown into less dangerous forms as a result of the immobilization. These interactions may be for long term (over decades), medium-term (over months) or short term (days). The soil constitutes the vertical sequence of layers or horizons which is differentiated by the physical, chemical, or biological characteristics.

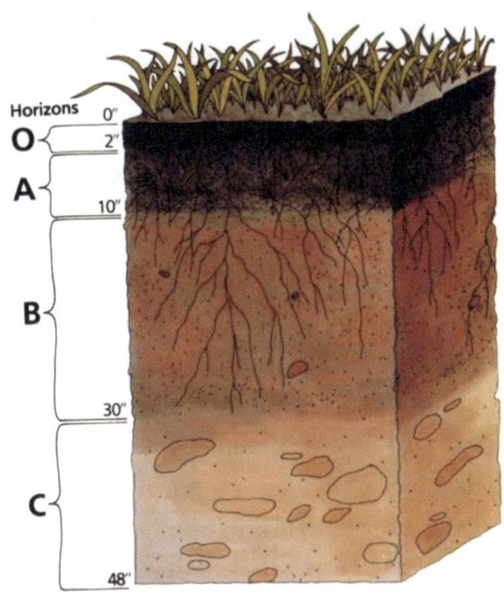

Figure 6.1: Horizons of soil.

Source: Image by Wikipedia.

The sequence of layer is known as the soil profile. The accumulations of organic materials as a result of the addition of plant and animal residues is the characteristics of the surface horizons. The surface horizons that contains a mixture of mineral and organic materials are identified as "A" horizons. It can be described by the amount of organic matter present within the horizon.

The surface layer that contains substantial proportion of organic material is the "O" horizon. The A and O horizons are subject to considerable changes as a result of the weather. Because it is frequently exposed to wetting drying cycles and to a wide range of temperatures. This results in the high rates of activity in physical and chemical processes. These are generally considered as the most active part of the soil biologically.

There are many functions being performed by the soil. The functions are related to the natural ecosystems, agricultural productivity, environmental quality, soil as source of raw materials and as base for buildings.

The most widely known function of the soil is the agricultural productivity. As per the agricultural productivity function, function of soil is to support plant and animal productivity whilst maintaining or enhancing water and air quality and supporting human health and habitation.

The soil should provide a physical, chemical, and biological context. This is for the suitability for the survival and development of living organisms, the conditions for the regulation and partitioning of water flow, storage, and recycling of nutrients and other elements; conditions to support biological activity and diversity for plant growth and animal productivity; an environment to filter, buffer, degrade, immobilize, and detoxify organic and inorganic substances; and provide mechanical support for living organisms and their structures.

The mentioned list of requirements implies the complex demand placed upon the soil if it is to maintain the productivity which is required for meeting the ever-growing demand of the ever-growing population. Further, there is a need to understand the complex interaction between the various components within the soil and the interactions with other components of the terrestrial ecosystems.

The process and the components of the soil needs to be investigated and understood in order to perform these varied functions effectively. This function is to be performed with no reduction in the quality of the soil. It must be ensured that one function of the soil must not reduce the ability of the soil to perform the other function.

Since most of the human activities directly or indirectly depend on the soil, it is considered to be essential for all the human activities. The environment affects the soil in a way that leads to soil erosion.

6.2. SOIL EROSION

Soil erosion is a widespread global issue that adversely affects the productivity of the quality of the soil and the natural ecosystem as well (Lal and Stewart, 1990; Pimentel, 1993; Pimentel et al., 1995; Pimentel and Kounang, 1998).

> Soil erosion is the process by which soil primary particles and aggregates are removed and lost from their point of origin by wind or water, or even mass wasting from gravitational forces (Troeh et al., 2004).

Figure 6.2: Soil erosion.

Source: Image by Wikimedia Commons.

With the increasing population, the requirement for ever-increasing food supplies will never end. More than 99.7% of human food (calories) comes from the land (FAO, 1998), while less than 0.3% comes from oceans and other aquatic ecosystems. The quality and productivity of the soil is the basic matter that affects the world food supply.

The man-made soil erosion over many years are significant and it has turned the valuable land to unproductive land and often eventually abandoned (Pimentel et al., 1995; Young, 1998). In other words, soil erosion diminishes the soil quality and thereafter reduces the productivity of natural, agricultural, and forest ecosystems (Pimentel and Kounang, 1998; Pimentel, 2001).

6.3. CAUSES OF SOIL EROSION

The main reason of the soil erosion is the exposer of soil to rain and wind. There is a great energy in the raindrops that hits the soil and it easily affects the soil particles. This way the thin layer of the soil gets removed from the land surface and it then creates a term known as sheet erosion. This erosion is the dominant form of soil degradation (Troeh et al., 1991; Oldeman, 1997).

The impact of the soil erosion is more on the sloping land, where more of the soil is taken away by the water that splashes down into valley or waterways. Wind energy has great power to take away the surface soil particles to a greater distance.

6.3.1. Soil Structure

Soil structure is the one that ease the process of soil erosion. The most easily eroded soil is the one that has medium to fine texture, low organic matter content, and weak structural development (Bajracharya and Lal, 1992). Basically, this soil type has low water infiltration rates and hence it has high rates of water erosion and the soil particles are easily displaced by wind energy.

Figure 6.3: Granular form of soil structure.

Source: Image by Flickr.

6.3.2. The Role of Vegetative Cover

The land areas that is covered by the plant biomass, living or dead, are more protected and experience relatively little soil erosion. This is because the raindrop and the wind energy are dissipated by the biomass layer and the topsoil is held by the biomass (Agriculture California, 2002; SWAG, 2002).

Figure 6.4: The vegetation cover either living or dead reduces the chances of soil erosion.

Source: Image by Flickr.

The extensive soil erosion is done by the removal of forests for crops and pastures. In developing countries, there is a widespread loss of soil vegetative cover where the population size is large. The agricultural practices at these places are often inadequate to protect the soil erosion.

6.3.3. Land Topography

The susceptibility to erosion of the soil is determined by the topography of a given landscape, its rainfall and/or wind and exposure.

In the country like the Philippines, where more than 58% of the land has a slope of greater than 11%, and in Jamaica, where 52% of the land has a slope greater than 20%, soil erosion rates as high as 400 t/ha-yr have been reported (Lal and Stewart, 1990).

The marginal and steep lands have a high rate of erosion which have been converted from forests to agriculture to replace the already eroded, unproductive cropland (Lal and Stewart, 1990).

6.3.4. Other Soil Disturbances

About three-quarters of the soil erosion worldwide is the world agricultural production. Soil erosion occurs when humans remove the vegetative cover (Lal and Stewart, 1990; FAO, 2002). The construction of roads, parking lots, and buildings are the examples of the issue.

Figure 6.5: Road construction results in soil erosion.

Source: Image by Geograph.ie.

Once the grass or the weeds are grown to the land surface naturally, the problem of erosion decreases (IECA, 1991). However, if the soil is covered by the parking, roads, building, the land cannot be used for vegetation production and hence the water runoff in adjacent areas increases.

This is evident along the stream banks, where erosion occurs naturally. Erosion occurs from the powerful action of adjacent moving water. There is an increase in the soil loss that occurs on steep slopes (30% or more), when a stream cuts through adjacent land. Also, on relatively flat land having only 2% slope, stream banks are eroded, especially during heavy rains and flooding. Further, the presence of cattle in and around streams, increases the streambank erosion.

The erosion impact from earthquakes is comparatively less as these natural calamities occurs rare. However, when earthquakes occur, heavy amounts of soil, that includes crops and forests are affected in area hillsides and in surrounding areas.

6.4. ASSESSING SOIL EROSION

There has been a significant impact of the soil erosion on the soil quality over the billions of years. However, the soil erosion has been taking place in the ecosystem slowly. There had been low rate of erosion that is sustained over billions of years, it has resulted in the displacement of enormous quantities of soil.

6.5. LOSS OF PRODUCTIVITY IN MANAGED ECOSYSTEMS

The total land on the earth has 50% of its land surface devoted to agriculture, and of this, about one-third is planted to crops and two-thirds to grazing lands (USDA, 2001). About 20% of the land area is occupied by forests (WRI, 1997). The cropland is more susceptible to erosion because of the frequent cultivation of the soils. This is because of the reason that before planting the crops, the vegetation is removed.

6.6. EFFECTS OF EROSION ON TERRESTRIAL ECOSYSTEMS

There is an adverse impact on the productivity of the terrestrial ecosystem. Soil erosion results in the increase in the water runoff that decreases the water infiltration and the capacity of the soil to store the water (Troeh et al., 1991; Pimentel et al., 1995; Jones et al., 1997). The soil erosion is harmful as the organic matter and essential plant nutrients are removed from the soil and the soil depth is reduced.

The changes in the quality of the soil does not only inhibit the vegetative growth but also reduce the presence of the valuable biota and the overall diversity of the soil (Troeh et al., 1991; Pimentel et al., 1995).

6.7. BIOMASS AND BIODIVERSITY

There is a direct relationship between the biological diversity existing in any ecosystem and the amount of living and non-living organic matter present in the ecosystem (Wright, 1990; Heywood, 1995; Lazaroff, 2001; Walsh and Rowe, 2001; Wardle et al., 2004). Erosion reduces the overall biomass and productivity if there is diminishing soil organic matter and soil quality.

Figure 6.6: Soil erosion has affected the biodiversity.

Source: Image by Wikimedia Commons.

At the end, soil erosion affects the plants diversity, animals, microbes present in the entire ecosystem. There are numerous positive associations that have been established between biomass abundance and species diversity (Elton, 1927; Odum, 1978; Sugden and Rands, 1990; Giampietro, 1997, PC, Insituto Nazionale della Nutrizione, Rome, Italy).

6.8. SEDIMENTS AND WIND-BLOWN SOIL PARTICLES

The impact of the soil erosion reaches to the surrounding environment that is beyond the damages to the rainfed agricultural and forestry ecosystem (Gray and Leiser, 1989; FEMAT, 1993; Ziemer, 1998). For example, there are large amount of eroded soil that is deposited in the streams, lakes, and other ecosystem.

In some areas, there is heavy sedimentation that leads to the flooding of river and lake (Myers, 1993). For example, the flooding occurred in the Midwestern United States during the summer of 1993 was caused by increased sediment deposition in the Mississippi and Missouri Rivers and their tributaries. The deposits raised the waterways, making them more prone to over flowing and flooding (Allen, 1994).

The water gets contaminated due to the sediments and harm the aquatic ecosystem. The sediments disrupt contaminate the water with soil particles and the fertilizer and the pesticide chemical they contain (Clark, 1987). The siltation of reservoirs and dams results in the reduction of the storage capacity of the water in dams and thereby increasing the cost of the dams and reduces the lifetime of the reservoirs (Pimentel et al., 1995).

6.9. ECOLOGICAL AND POLITICAL ASPECTS OF SOIL PROTECTION

Other environmental compartments (water and air) are strongly cross-linked to soil. The interaction with the other environmental compartments exists because of these linkages. It is due to this, that soil have an important ecological mediator function. This is with respect to the material transport processes, that soils can be both sources and sinks.

Human has taken possession of the soil from the early times in contrast to the other environmental compartments (air, water systems). The concept of the "owned land" that means "possession of the soil" is based on this fact. Therefore, soil is the most important resource in human life for both the aspects material and cultural.

Through this, the usage of soil can be restricted which would have been used for conservation purposes or as a groundwater reservoir for production of drinking water. The conflicting interactions are more complicated with cross-linked and complex ecosystems. These are not transparent and easy to understand.

It should always be kept in mind that while exploiting the soil the benefit from the use of one function can be detrimental to the use of other function. Soil protection is a main task where the utilization of the soil function has to be selected prudently.

There must be a proper way of targeting the end-users and the management of these users should be in such a way that there is no way left for the conflicts among various intended users. The kind and the extent of the usage of the soil must be taken into account. The understanding among the users about the soil protection as the management of competing claims for usage by humans is no doubt an anthropocentric view.

The care should be done in such a way that the exploitation of the soil in the past should have no harmful effect on the present-day or future uses. Therefore, the selection of the appropriate user is an important task.

From the discussions, it can be said that soil is a mixture of minerals, water, air, and organisms, together with products of their transformation and degradation. There are different types of soil that can be formed depending on the climatic conditions and the nature of the underlying rocks. The rate of formation depends on the rocks whether it is weathered in the lower layers and on how rapidly biomass is degraded in the upper layers.

6.10. FUNCTION OF SOIL

The most important function of the soil are the regulating and productive functions, known together as ecological soil functions. The other soil functions are the raw material resources, land for building, recreational areas, and waste disposal. The mechanical retention of particles and colloids by the soil-pore system is the filter function of the soil.

On the other hand, the buffer function includes the binding of dissolved substances by absorption and desorption. It also includes the inclusion and release of substances that contain water by pore spaces, and the in- and ex corporation of substances by soil organisms.

The habitat function enables the soil to support colonies of organisms within and upon the soil. The function of the soil as substrate for higher plants is also there. There is a difference in the productive function and the habitat function where in the latter, economic exploitation of the biomass is of primary importance.

However, the other functions of the soil are of anthropogenic nature. In this, the ecological functions are of less importance as compared to the anthropogenic exploitation or are indirectly exploited. Regardless of the use to which it is put, the regulatory and habitat functions can be defined solely on the basis of the natural pedological and ecological properties of the soil.

The productive function is recognized on the basis of both ecological aspects (habitat for cultivated plants) and socio-economic criteria. This is because the productive function is directly interlinked with agriculture and forestry.

6.10.1. Soil Ecosystems

Pedogenesis is the formation of soil that involves interactions between the lithosphere, hydrosphere, atmosphere, and biosphere. The weathering forces and organisms convert the lithosphere into the pedosphere at the earth's surface.

Various soil-forming factors results in the formation of the characteristic horizontal layers, known as horizons. The biological, chemical, and physical properties of the horizons are different. The horizons have the large proportion of organic material at earth's surface. However, the mineral component increases with the increasing depth.

The horizons together make up the solum. The main criterion for the classification of soils is the vertical distribution of the horizons in a two-dimensional cross-section through the soil profile. There is a term called as pedon, which is the smallest and the undividable unit of the soil body. There exists a peda which is from the same soil type and forms a polypedon. The properties of both are slightly different in terms of the horizon depth.

Lithosphere

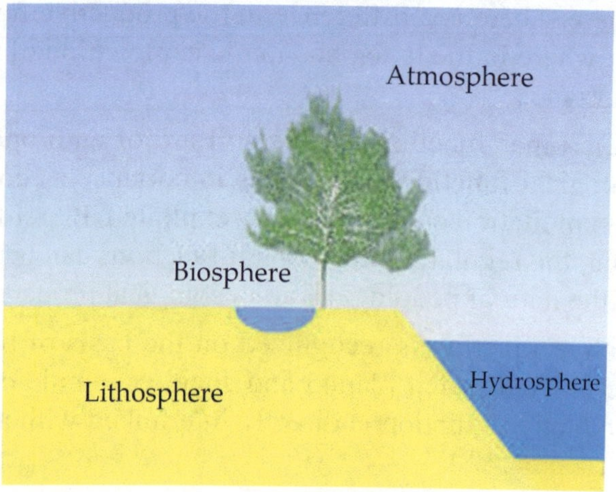

Figure 6.7: Lithosphere.

Source: Image by Wikimedia Commons.

The lithosphere is located beneath the ocean. It is the rock mantle of the continents and the earth's surface. The upper part of the soil is formed by the rock weathering, mineralization, decomposition, humification, texture formation, and displacement, under certain conditions (e.g., climate, leaf fall from vegetation).

The chemical, physical, and biological transformation of inorganic material such as basalt, granite, gneiss, carbonate, loess, and sand at the earth's surface is the weathering. It is done by size reduction, degradation, transformation, and dissolution. The environment impacts the soil development that the petrogenic starting materials are increasingly replaced by pedogenic materials such as clay minerals, which are more stable at the earth's surface.

6.11. MAJOR SOIL PROCESSES

6.11.1. Decomposition: Immobilization and Mineralization

More than 90% of the terrestrial net primary production (NPP) is returned as a dead organic litter. This litter consists of leaves, roots, and wood from trees and organic residues from agricultural fields. This is decomposed in the soil, and the nutrients contained within it is recycled for further use.

The complex food web is driven by the decomposition process. The numerous interactions between the initial agents of decomposition, the bacteria and fungi, and the fauna which hi turn feed on them.

6.11.2. Nitrogen Cycle: Major Processes

The other major process of the soil is the nitrogen cycle which enters the ecosystem via nitrogen fixation. Under this, the dinitrogen molecule (Nj) is separated into two nitrogen atoms. There is a considerable expenditure of energy and the assistance of the nitrogenase enzyme, to break the triple covalent bond.

6.12. LAND

There are numerous changes in the pattern of the land use. The total area of the cropland, pastureland, and rangeland decreased, while there has been an increase in the area of developed land.

6.12.1. Socioeconomic Impacts

Along with labor and capital, land is one of the major factors of production in the classical economics. This is considered as an important input for housing and food production. Therefore, for any agricultural economies, land use is the backbone and provides substantial economic and social benefits.

There is a cost attached to the use of the land. The farmland forests are converted into urban development that results in the reduction of the amount of land left for food grain production and timber production.

The environmental effect on the quality of the soil degrades it for any further use. Soil erosion, salinization, desertification, and other soil degradations associated with intensive agriculture and deforestation reduce the quality of land (Lubowski et al. 2006).

Soil erosion, salinization, desertification, and other soil degradations associated with intensive agriculture and deforestation reduce the quality of land resources and future agricultural productivity (Lubowski et al. 2006). There are many challenges for the farmers on the urban side. The people living near by the farms are more assumed to harm and destruct the crops and the equipment of the farms (Lisansky, 1986).

The farmers that lives nearby the farms are often cooperative in production activities. This includes equipment sharing, land renting, custom work, and irrigation system development. Also, it must be noted that if the nearby farms are converted into development then the help would also stop. Farmers would not be able to take any benefit from them.

There will be the environmental effect on the urbanization. This may also cause the "impermanence syndrome" means a lack of confidence in the stability and long-run profitability of farming. This will lead to reduction in investment in new technology or machinery, or idling of farmland (Lopez, Adelaja, and Andrews, 1988).

Due to urbanization, there exists important opportunities for the farmers. Farmers can sell higher value crops to the new customer base emerged. There has been changes in the rural communities in many places. There have been increase in the lower income group and decrease in the upper income group in the cities.

The urban and rural people's problem are solved through suburbanization. Most of the land areas are rural, most watersheds are in rural places, and most of the atmosphere exists above the rural space. Rural people are concerned about the urban pressure on the rural land and other resources.

6.12.2. Environmental Impacts

The environmental impact on the land use change is the most pervasive socioeconomic force driving changes and degradation of ecosystems. There is a substantial change in the earth's surface that includes, deforestation, urban development, agriculture, and other human activities. Such disturbance of the land affects the important ecosystem processes and services. This could have wide-ranging and long-term consequences.

There is valuable habitat and open space for many wildlife species. However, there is severe ecosystem of the intensive agriculture consequences. It has been recognized that agricultural land use and practices can cause water pollution and the effect is influenced by government policies.

Water runs off from the agricultural land and is the leading source of water pollution both in inland and coastal waters. Many species are brought to the verge of extinction by the conversion of the wetlands to crop production and irrigation water diversion.

There are several ecosystem services that the forests provide. They support biodiversity, providing critical habitat for wildlife, remove carbon dioxide from the atmosphere, intercept precipitation, slow down surface runoff, and reduce soil erosion and flooding.

When the forests are converted to agriculture or there is an urban development, the important ecosystem will be reduced or destroyed. The environmental impact on the agriculture, like deforestation has substantially altered and fragmented the earth's vegetative cover.

Such environmental disturbance can change the global atmospheric concentration of carbon dioxide, the principal heat-trapping gas, as well as affect local, regional, and global climate by changing the energy balance on earth's surface (Marland et al. 2003).

The development of the urban areas is linked to many environment problems that includes air pollution, water pollution, loss of wildlife habitat. While the urban runoff contains the nutrients, sediment, and toxic contaminants, and can cause not only water pollution but also large variation in stream flow and temperatures.

The leading causes of biodiversity decline and species extinctions are habitat destruction, fragmentation, and alteration associated with urban development (Czech, Krausman, and Devers 2000; Soulé 1991). Urban development and intensive agriculture in coastal areas and further inland are hazardous to the health, productivity, and biodiversity of the marine environment throughout the world.

6.12.3. Policy Implications

There are many economic and social benefits of land use. However, it carries a cost to the environment. Most of these costs are used in the land use decisions, while most of the environment externalities are not. These environmental "externalities" renders the private and social costs for some land uses, that finally leads to an inefficient land allocation.

There is an instance, that the developers may not bear all the environmental and infrastructural costs of their projects. There is an agricultural commodity and an open space that is provided by the farmland. Although farmers are paid for the commodities they produce. However, they are not compensated for the open space they provide. Therefore, the market prices of the farmlands are below their social values.

There is a need for private conservation efforts and public land use planning and regulation for such market failures. There are different forms a land use regulation can take. The most traditional command involves zoning, density regulation, and other direct land use controls. These policies can be used as regulatory tools, as they could lead to substantial social welfare loss. The loss will be in the form of higher housing prices, smaller houses, and inefficient land use patterns (Cheshire and Sheppard 2002; Walsh 2007).

In many communities, there is a contentious issue for those who are facing rapid urbanization. The land use planning helps in protecting the farmland, water quality, open space, wildlife habitat and also increases the value of the property and health.

The critics of land use regulation argues that the urban development is an orderly market process that allocates the land for the urban use. These lands rarely bear the cost of regulation and that is why the government over regulates it.

6.13. LAND AND MAN

Although man came after the life started on the earth, the man is treated as a person who can change the system of the earth. The technology has helped humans a lot, there is a continuous effort in living better life. It can be attained with the help of identification and utilization of the means given by the physical environment.

Land is the most important component of the system of the earth. It is the immediate victim of all sorts of the human activities. Worldwide, with the increasing population, land is in demand for fulfilling the basic needs. The needs are not only related to the increasing demand for food and space but for better material expectation also.

In the last few years, there has been human intervention of land resources. The co-existence of biotic and abiotic elements of the earth system makes it live and unique, as earth is the living space for all the species. Also, man cannot sustain alone on the earth. The main concern is the exploitation and the unplanned use of land.

Land is a resource base by itself. It is the place over which the living species live. However, man has achieved many advancements on the technology front, even then the man is a terrestrial animal. Man has become the aster in the utilization of the earth's landscape. The technological and socio-cultural growth of human society leaves its impact on the land more than any other resource base.

However, there has been an estimate that more than two-thirds of this land surface is not suitable for human habitation. There are areas where there is less population due to the extreme climatic conditions and inhospitable terrain. The environmental conditions had rendered the land ineffective for the use by man due to the wet climate Also, it is estimated that only 21% of land area is dry and desertic.

6.14. HOLISTIC CONCEPT OF LAND

The physical factors like geology, topography, hydrology, soils, microclimates, and flora and fauna has played an important role. These are constantly on interaction under the influence of man's interference. This makes the land a holistic entity. Therefore, land has become the combination of physical as well as socio-cultural factors.

As per FAO, the land is "a delineable area of the earth's terrestrial surface, that encompasses all the attributes of the biosphere immediately above or below this surface, including those of the near-surface climate, the soil and terrain forms, the surface hydrology (including shallow lakes, rivers, marshes, and swamps), the near-surface sedimentary layers and associated groundwater reserve, the plant and animal populations, the human settlement pattern and physical results of past and present human activity (terracing, water storage or drainage structures, roads, buildings, etc.).

In terms of its topography and spatial nature, the land is a physical entity. The land is often associated with an economic value, and the value is expressed in price per hectare at ownership transfer" (FAO, 1995). According to Stewart (1968) the term land is used in a "comprehensive, integrating sense. This is to refer to a wide array of natural resource attributes in a profile from the atmosphere above the surface down to some meters below the land surface.

There are some natural resources attributes such as climate, landform, soil, vegetation, fauna, and water.

The main natural resource attributes are climate, landform, soil, vegetation, fauna, and water." Hoover and Giarratani (1984) state that land, first, and foremost denotes space.

Land is to be understood from different perspective and dimensions. Therefore, land is a fundamental resource. Different disciplines have investigated land due to its varying dimensions. The basis for a human life is land and supports other life forms too. There are other resources which are useful for the human society.

6.14.1. Land as a Biological Entity

Land is the basis for the terrestrial biodiversity. This resource base consists of all sort of the biological habitats. Different types of flora and fauna lives on the la nans grows and interact with each other. Therefore, land is considered as the great support system for all the living organism and the terrestrial life.

The environmental condition makes it possible for land to carry on the functions. Land produces biomass for the formation of the biotic materials. The basic needs of the human are fulfilled by the land, i.e., food, fodder, fiber, fuel, timber, and other biotic materials. The needs are fulfilled either directly or through animal husbandry that includes aquaculture and inland and coastal fishery.

Therefore, land can be viewed as the habitat for various life forms. This approach is mainly followed by biologists and considers land as a supporting base for living organisms.

6.14.2. Land as an Economic Entity

From the economic point of view, land is a resource in itself. Earlier, in the traditional rural societies, people considered land as a common good, a gift from God that permits the satisfaction of primary needs for food and shelter for all members of the group. While in the modern societies, land is considered to be the most precious economic resource.

People draw most of the sustenance and the fuel, clothing, and shelter from the land itself. As an economic view, land is the fundamental place to live and work. Land is used for the productive economic factor in agriculture, forestry, grazing, fishing, and mining. Therefore, in the economic forces, land has become the inevitable factor of production.

However, after the production, there are various other activities that is carries over the land. Additionally, land is considered as a consumption good, property, and capital. This is the basis of social prestige, wealth, and political power.

6.14.3. Land as a Physical Entity

Land is also the physical basis of cultural growth. Land is the solid crust of the earth that is composed of different materials. There exists lithosphere on the earth, which is made up of the physical and chemical characteristics.

There are natural irregularities, landforms, which make the physical set up of the land so complex and hence, the structure of the land is so uneven.

The higher order landforms constitute mountains, plains, and plateaus. They are formed as a result of tectonic movements of the earth. Since the land surface is exposed to atmospheric mechanisms, minor landforms are sculptured over it by gradational agents. Land is the victim of gradational work of rivers, winds, glaciers, sea waves and underground water.

The nature of geologic structure and gradational process of the land decides the regional topography of the land. The hydrological properties of the land are also to be noted. Water on earth is an elixir of life that makes the earth a lively planet. The life on earth is possible due to the hydrological cycle where there is transfer of water through hydrosphere, atmosphere, and lithosphere.

The role of land in the hydrological cycle is of great significance. Land regulates the storage and flow of surface and the groundwater resources. The hazardous compounds present in the water in the land is converted into the quality product and that is possible due to the receptive, filtering, and transforming function.

Thus, land is considered as a physical entity by the physical science by concentrating on its structure, form, and composition.

6.14.4. Land as a Spatial Entity

Fundamentally, space is the first thing that land occupies. With which it extends over an area. There is a varied combination of physical, socio-cultural, and economic parameters. Therefore, land has a different permutation combination arrangement over the space. Land is the matter of the advancement and the sustenance of the human society.

The area differentiation and spatial differentiation is the basic domain for the purpose of research from the view of geography. Therefore, from the viewpoint of the geographers, the land is the foundation on which man has organized the life in different forms of society. Land is a scarce resource, the demand of which is increasing day by day with the growing population.

The population growth, limited expansion of arable land and the need for land for non-agricultural purposes has altogether increased the pressure on the competition on land. Other than land, there is no other resource which is put underuse, misuse, abuse or overuse.

There is a continuous threat to the land resource because of the variety of natural degradation processes and mismanagement of land by humans. There is a need for scientific and judicious use of every piece of land. There must be effective tackling of the existing land resource. The effective use of land resource can be achieved by developing various strategies for optimal and sustainable use of this finite resource.

Consequently, there has been a significant change in the land use. To meet the increasing demand of the growing population, it is important to optimally utilize the land resource for the selection, planning, and implementation of land use schemes.

6.15. CONCEPT OF LAND USE

The utilization of the surface of all the vacant and the developed land on a specific point at a given time and space is known as the land use (Mandal, 1990). As per the FAO concept, land use is the human activity directly related to the land, making use of its resources and impacting directly on the land.

The main emphasis is on the function or the purpose for which the land is used and there the reference is made to the "the management of land to meet human needs" (FAO, 1976). The land use of an area is "the cumulative outcome of the historical events, the interaction of the economic forces with the natural environment and natural value of the society" Jasbir Singh and Dillon (2005).

Therefore, land use involves the concepts of optimizing the land use potential, land evaluation for example, and of land use planning. There must be a distinction between the present land use and the potential land use.

6.16. FACTORS OF LAND USE

The natural and socio-economic factors which decide the utilization of land by man over time and space is the land use pattern. The land is made up of four variables, land, water, air, and man. Every variable has a role to play in the composition of the life.

It is dependent on the man to efficiently utilize the land and to manage it. According to the purpose of the land use, there is an intensity and varied nature it serves. It can be food production, recreation, or mining- and the biophysical characteristics of the land itself. The use of the land is intensified with the increase in the population, method, and technology.

It is the farmers, owners, institutions of government who decides as to what the land would be used for according to the perception of the needs (Mandal, 1990). These decisions are taken after considering large number of factors of physical and cultural aspects.

6.16.1. Physical Factors

There are some environmental impacts on the land use. The first is the physical factors that controls the land use. It includes mainly geology, topography, climate, and soil.

1. Geology

There is a direct and indirect influence of the geological structure on the land use. Over a region, the nature and composition of bedrock and distribution of mineral deposits directly influences the type of land use.

The indirect controls that affects the land use are the process of soil formation, drainage pattern, water table, structure, and texture of regolith, etc.

2. Topography

As per the topography, there is a different type of land use. Where the country is mountainous, hills, and valleys are the factors that has to be taken into notice while using the land. In the region where the surface is rugged, the agricultural practice cannot occur while in the case of ideal slope terraced farming can be practiced.

Many mountainous regions are covered with the different kind of density and type of crops which differs as per the amount of rainfall it experiences. The erosional surface has less scope of agricultural practice because of the lack of upper soil cover. These plateaus have flat top surface. The valleys are also very deep to be suitable for the agricultural practices.

Therefore, plains are considered to be the ideal place for the agricultural practice. As about 75% of the world's population lives on the plain region.

3. Climate

Climate is a dominant factor that controls the use of land. It controls the agricultural land use. For good agricultural production, there is a need of favorable climatic conditions.

There is always a need of a favorable climatic conditions. People usually lives away from the places where there is extreme climatic condition. The elements of climate like temperature, rainfall, humidity, etc. has an important role to play in the plant growth and human activities as well.

4. Soil

The nature and distribution of different types of soil is also an important factor of land use.

6.16.2. Cultural Factors

The cultural factors of land use include demographic, social, and institutional aspects of the society.

1. Demographic

While deciding on the land use, the distribution, growth, and density of population in a particular area are to be taken care. There is a raising conflict in the land use due to the growing population and the pressure on the land. The rural urban character of the people of the region is an important factor of the land use.

2. Social

The social setup of a region is also considered as a dominant controlling factor of land use. The factors important here are social customs and beliefs of the people, land tenure system, size of landholding, social security, stand of political parties, etc. The crucial part in it is the level of education and technology.

3. Institutional

Institutional factors like government policy, constitution, legislation, stand of political parties, tax regulations, system of civil service, tariff control, etc. are the factors for land use.

6.17. APPROACHES TO LAND USE STUDIES

Land is a multi-faceted entity. Mandal (1990) has suggested some of the approaches for the study of land use as given below.

6.17.1. Genetic Approach

This is related to the study of the genesis of the land use pattern. The historical evolution of the cultural elements of the landscape is studied for the estimation to be made for each and every plot of the land.

6.17.2. Prescriptive Approach

The better utilization of the limited land for the welfare of the human society, present, and future is the prescriptive approach.

6.17.3. Remedial Approach

Under the remedial approach, a study is conducted for searching the better use of land by suggesting the removal of a certain inferior use.

6.17.4. Case Study Approach

There is a detailed micro-level study of land utilization characteristics of the selected study area.

6.17.5. Regional Approach

Under this approach, the region is considered as a whole. The influence of regional geographical factors on various land uses is analyzed in this.

6.17.6. Ecosystem Approach

Ecosystem approach deals with the association of environmental factors that affect the distribution of various land uses.

6.17.7. Statistical Approach

Based on this approach, the quantitative analysis is used. Inferences and conclusions are derived by the quantitative analysis of land use statistics.

6.17.8. Principle Approach

There are certain models and principles that considers the pattern of land use of a particular area. The approaches put forwarded by Von Thunen, 1826 and Marx are examples.

6.18. LAND USE AND LAND COVER

The terms land use and land cover are used interchangeably. They can get overlapped because the distinction between the two cannot always be different. Land use is the activity of the man on the land which is directly related to the land. Whereas, land cover is the "the vegetational and artificial constructions covering the land surface" (Burley, 1961).

The biophysical state of the earth's surface and immediate subsurface is the land cover (Turner et al., 1994, 1995). In other words, "describes the physical state of the land surface: as in cropland, mountains, or forests" (Meyer et. al 1992, 1994, cited in Moser 1996). Meyer and Turner (1994, 1996) add: "it embraces, for example, the quantity and type of surface vegetation, water, and earth materials."

Moser (1996) described: "The term originally referred to the type of vegetation that covered the land surface, but has broadened subsequently to include human structures, such as buildings or 63 pavement, and other aspects of the physical environment, such as soils, biodiversity, and surfaces and groundwater."

According to Meyer (1995), "land use is the way in which and the purpose for which, human beings employ the land and its resources." "Land use involves both the manner in which the biophysical attributes of the land are manipulated and the intent underlying that manipulation -the purpose for which the land is used" (Turner et al., 1995).

Land use has been described briefly, "denotes the human employment of land" (Turner and Meyer 1994). According to Skole (1994) "Land use itself is the human employment of a landcover type, the means by which human activity appropriates the results of MSWMNPP as determined by a complex of socio-economic factors."

FAO (1995, 1996) denotes that "land use concerns the function or purpose for which the land is used by the local human population and can be defined as the human activities which are directly related to land, making use of its resources or having an impact on them." Doi (2007) has referred the land use to the human activity or economic function associated with a specific piece of land. The land cover is the type of feature actually present on the surface of the earth.

The intentional role of the people to adapt the natural land cover for their benefit differs the land use from the land cover.

6.19. LAND DEGRADATION IN GLOBAL CONTEXT

The environmental impact on the land is the land degradation. It is a global problem which is associated with desertification in arid, semiarid, and dry sub-humid zones, commonly called the 'dry-lands.' Even now, land degradation is the contested topic in its determinants, degree, distribution, and effects.

Figure 6.8: Land degradation.

Source: Image by Flickr.

As per UNEP 1997, the dry lands which is almost 47% of the total globe's surface, the population is long associated with lands. The land degradation is considered to be variable, discontinuous, arising from different causes and affecting people differentially according to their economic, social, and political circumstances (e.g., Mortimore, 1998).

According to some sources, there are up to 70% drylands, that are desertified. However, the percentage is not clear yet some other says that it is only 17% (Reynolds et al., 2003). In the midst of such discrepancies, it is difficult to identify that how the land degradation is a global concern.

Adams and Eswaran (2000) has estimated a number that 26 billion people in more than a hundred countries are covering over 33% of the earth's land surface. The rangelands of about 73% in drylands are currently being degraded, along with 47% of marginal rainfed croplands and a significant percentage of irrigated croplands.

There is an open debate regarding the science behind the actual extent of land degradation (Stocking, 1996). Economically and geographically, drylands are the home of the world's poorest and most marginalized people (Dobie, 2001).

6.20. LAND DEGRADATION CONTROL AT GLOBAL LEVEL

In order to address the land degradation and the problems related to it that it brings for the poor developing countries, there is a main global agreement for this issue and that is, The United Nations Convention to Combat Desertification (UNCCD). This initially started in the back 1977 in the United Nations Conference on Desertification. They adopted a plan of action to combat desertification.

However, the plan was ineffective and largely ignored. It had a little support from the developed or developing countries, and even less from potential donors. The United Nations Environment Programme (UNEP) has setup a global consultation to ensure that the land degradation would be included in the next major environmental conference.

6.21. CONCLUSION

Soil is the upper layer of the earth in which the plants grow. Soil is composed of minerals, soil organic matter, living organisms, gas, and water. Soil is an important medium for plant growth, and it support the animal and human activity. Soil sometimes acts as a reservoir for nutrients and water providing the plants' needs for these requirements throughout their growth. However, the environmental impact on soil adversely affects the quality of the soil. It leads to soil erosion. Soil erosion is the loss of the soil from the land surface. Throughout the world, the growing population has increased the tension of population due to soil erosion, water availability, loss of biodiversity. The changes by humans on the land has significantly resulted in the valuable land becoming the unproductive land.

Land degradation has the similar adverse impact on the living organism, plants, animals, and other habitat. It is, therefore, largely a local problem and there is a need of local solutions, such as better soil conservation and attention to local policies. The land degradation has adversely impacted the agricultural production as the quality of the land diminishes. There is a pressure of the growing demand for the food and other agricultural produce of the increasing population. However, there is again a growing concern and it is deforestation. Concluding the chapter, there must be some actions taken with respect of the soil erosion, land degradation, deforestation that has a direct impact on the soil and land for the agricultural productivity.

REFERENCES

1. Coleman, D., (n.d.). *Soil Biota, Soil System and Processes*. [eBook] Available at: http://coweeta.uga.edu/publications/1222.pdf (accessed on 4 March 2020).
2. Gisladottir, G., & Stocking, M., (2005). Land degradation control and its global environmental benefits. *Land Degradation & Development*, *16*(2), 99–112.
3. Land Degradation, (n.d.). [eBook] Available at: https://na.unep.net/atlas/kenya/downloads/chapters/Kenya_Screen_Chapter4b.pdf (accessed on 4 March 2020).
4. Nortcliff, S., Hulpke, H., Bannick, C., Terytze, K., Knoop, G., Bredemeier, M., Schulte-Bisping, H., Auerswald, K., Litz, N., Mayer, R., Stoy, A., Alef, K., Kerndorff, H., Crössmann, G., Eikmann, T., Franzius, V., Grimsky, D., Möhlenbruch, N., Storm, P. and Dworshak, U., 2001. Soil. *Ullmann's Encyclopedia of Industrial Chemistry*, [online] p.1. Available at: <https://onlinelibrary.wiley.com/doi/abs/10.1002/14356007.b07_613> [Accessed 4 March 2020].
5. Soil Erosion: A Food and Environmental Threat, (2005). [eBook] Available at: http://saveoursoils.com/userfiles/downloads/1368007451-Soil%20Erosion-David%20Pimentel.pdf (accessed on 4 March 2020).
6. Wu, J., (n.d.). *Land Use Changes: Economic, Social, and Environmental Impacts*. [online] Choicesmagazine.org. Available at: http://www.choicesmagazine.org/magazine/article.php?article=49 (accessed on 4 March 2020).

CHAPTER 7

Impacts on the Biological Environment

CONTENTS

The biological environment includes all the living things on the earth. It includes man, plants, and animals. Since man is the most important component of the biological environment, it is therefore necessary to preserve the life on earth. This chapter discusses all about the biological diversity, ecosystem, loss of biodiversity in the first half. The mid of the chapter gives insights about the wetland ecosystem, its types, and classification. There is a small topic that explains about the sea as the biological environment and the biological impact assessment process. The ending of the chapter has topics related to ecosystem services and the conservation of flora and fauna.

7.1. ECOSYSTEM

A region that has forest, grassland, desert, wetland or coastal area around is what constitutes an ecosystem. The area has specified and recognizable landscape which forms the ecosystem. The ecosystem has a wide variety and it has different nature too. The nature of the ecosystem is based on its geographical features such as hills, mountains, plains, rivers, lakes, coastal areas or islands.

The climatic conditions also play a vital role in controlling the ecosystem. The climatic changes may be such as the amount of sunlight, the temperature and the rainfall in the region. The non-living component is formed by its geographical, climatic, and soil characteristics.

Figure 7.1: Ecosystem on the verge of extinction.

Source: Image by Needpix.

Certain conditions are formed that helps the ecosystem to support a community of plant and animal. There is a biotic component referred to as the living part of the ecosystem. The ecosystem is divided into terrestrial or land-based ecosystems, and aquatic ecosystems in water.

All these help in forming two major conditions for the living organisms to live in the earth's atmosphere. The living organisms present on the earth lives in the communities of plants and animals. At different point of time, the living organism interact with the environment and with each other for large number of reasons.

However, the life on earth can be there only in the small proportion of the earth's land, water, and its atmosphere. Globally, at the skin of the earth, or the land surface, the sea, and air forms the biosphere.

The living community of plants and animals in any area together with the non-living components of the environment such as soil, air and water, constitute the ecosystem

7.2. BIOLOGICAL DIVERSITY

Biodiversity is basically the variety and variability of life on earth. Within the natural system, the variety and variation form a comprehensive umbrella that refers to the term biodiversity. Biodiversity is often understood in terms of the wide variety of plants, animals, and microorganisms, the genes they contain and the ecosystem they form.

Figure 7.2: Biodiversity, variety of life on earth.

Source: Image by Wikipedia.

The biodiversity present in the today's scenario is the result of billions of years of evolution. It is shaped by natural processes and, increasingly, by the influence of humans. Biodiversity is a web of life of which all the living and non-living organism are an integral part. The life of each and every species depends totally on the biodiversity.

According to the scientists, there are about 13 million species. However, the data relating to this as per UNEP is different. As per UNEP, the estimates are 9.0 to 52 million species exist on earth (Mora et al., 2011). Therefore, there are about 2.1 million species have been identified, mostly small creatures such as insects.

In biodiversity, there is genetic difference within each species- for example, between varieties of crops and breeds of livestock. The uniqueness of each and every individual is determined by the DNA, which is the building block of life, chromosomes, and the genes. There is another feature of biodiversity which is the variety of ecosystems such as those that occur in deserts, forests, wetlands, mountains, lakes, rivers, and agricultural landscapes.

In each ecosystem, humans, and other living creatures form a community where they interact with each other, air, water, soil around them. Therefore, biodiversity is considered at 3 major levels:

7.2.1. Genetic Diversity

This is a variety of genetic information contained in all of the individual plants, animals, and microorganisms which is present within the populations of species. Basically, it is the variation of genes within species and populations.

7.2.2. Species Diversity

This is basically the variety of species or the living organisms. The species diversity is measured in terms of richness of the species. This refers to the total count of species in a defined area.

7.2.3. Species Abundance

Species abundance refers to the relative numbers among different species. In case the species have the equal abundance level, this would refer that the variation is high that results in high diversity. However, in case of low diversity, the one species is represented by 96 individuals, whilst the rest are represented by 1 species each.

In nature, there is no species that is equally different. Therefore, it is possible to identify the species on the basis of their functions:

1. Functional Types

Those species which perform different ecological functions.

2. Functional Analogs

Functional analogs represent those distinct taxa who performs the same or very similar ecological functions.

7.3. LOSS OF BIODIVERSITY

The loss of biodiversity is the extinction of plants and animals worldwide. In human history, the loss of biodiversity and the related changes in the environment are now faster than ever before. There is no scope of the process getting slow down.

However, humans had already distorted and altered the natural resources of the earth. The earth's ecosystem has been used mostly for agricultural processes and other commercial uses. The population of plants and animals are diminishing at a faster scale.

However, any species whether related to humans or plants or animals, is said to be extinct on the earth one day or the other. The process of extinction is natural. But the human activity has increased the extinction rate by 100 times as compared to the natural loss. The range of driver is the reason of loss of biodiversity.

Any natural or human-induced factor that directly or indirectly causes a change in an ecosystem Is the driver. A direct driver influences the ecosystem processes. There are some direct drivers that affects the biodiversity, these are: habitat alteration, climate change, invasive species overexploitation and pollution.

7.4. THE SEA AS A BIOLOGICAL ENVIRONMENT

The chemical and physical aspects of the elements that together constitute the inorganic marine environment are:

- The seawater itself; and
- The ocean floors.

The decisive factors in the history and fate of a perplexing array of living things are the chemical constituents and the physical properties of the seawater, together with their distribution, concentrations, and cyclic changes, the movement of the water, and the nature of the ocean floors.

Figure 7.3: Biological environment of the sea.

Source: Image by Wikipedia.

There are many secrets of racial development that are held within. Also, the understanding of the delicately balanced maintenance of life and of the potentialities of future development must be sought out. Marine organisms are the part of the sea.

There are various organisms that live in the seawater and they are bound to the sea for their existence and are the part of the sea both racially and individually. This is just like there are two types of sea salt which is conservative and nonconservative (biologically changed).

The organisms like salts, are subject to the natural laws of the sea and are a part of the perpetual cycle of inorganic and organic substances. This is also important in many aspects of oceanography.

In general, the organisms have not developed highly specialized integuments and regulatory systems to protect themselves against sudden and intense environmental changes, so have most land animal. This is due to the stability of the physical characteristics of the seawater and of the composition and concentration of the dissolved salts.

Also, the small changes in the aquatic medium are promptly brought to play upon the population. Being a part of the dynamic environment, the organisms themselves modify mainly its chemical character by withdrawing or adding substances associated with the activities of life.

7.4.1. Physical and Chemical Characteristics of the Marine Environment

Water is the essential part in the maintenance of all the life on the earth. There is 80% or more by weight of active protoplasm. It is the most efficient solvent among all other solvent and carries in solution the necessary gases, oxygen, and carbon dioxide, as well as the mineral substances necessary to the growth of plants and animals. Therefore, it is itself one of the essential raw materials in the manufacture of foods by plants.

Organisms living in the terrestrial environment have devised means, such as impervious integuments, to conserve water, and the land plants have roots and special vascular systems for transport of water to all growing parts. In the marine environment, there is freedom from desiccation, except at high-tide levels. Therefore, there is no provision of highly specialized means for the conservation of water or for its transport in plants.

In terms of biological importance, heat capacity of water is high. The latent heat of evaporation makes the chances more obvious regarding the danger of rapid change of temperature in the environmental medium. It is also possible for the sea in sustaining the life of plants throughout a relatively deeper layer. This is due to the high degree of transparency of water. In animals, there has been the development of organs of vision and of orientation has progressed to a marked degree.

7.4.2. Other Characteristics of the Environment

There are certain biological properties and characteristic that is inherent in the seawater environment as a whole. This is apart from the chemical and physical properties.

This is the result of the magnitude of the ocean itself, its great depth, and its expense. When an ocean is considered as an entire environment, there is wide range of living conditions present. They may be interesting in the first go but these uniform conditions do not prevail all the time over the extensive areas.

Therefore, according to this, the faunal areas can be recognized which are characterized by some specific form.

On the other hand, in most of the restricted areas, a wide range of conditions may be encountered, especially in coastal regions. The physiographic character of the coastline, depth to bottom, topography, and nature of the bottom, inflow of land drainage, meteorological conditions, and so forth are the possible reasons for this.

The shallow depths and varying conditions are frequently favorable to abundant production of primary food. The gradients of salinity, light, and temperature that exist in the sea are favorable to a number of sensitive animals that possess the ability, through swimming or otherwise, to adjust themselves to optimum conditions and these must not be overlooked.

Depth and Light

There are number of important features of far reaching biological effect which is inherent in the vertical range or depth of the open sea habitat. The great vertical range of the euphotic zone which is available for production of floating microscopic plants is of utmost importance.

But the gradient of light, resulting from depth of water, both as to quantity and quality allows the adjustment of many animals to the optimum condition with respect to this factor. This is indeed associated with diurnal migrations of many forms to lighter or darker situations.

Pressure

The pressure from the abyssal region of the sea does not exclude life. But for water, it is little compressed, and the equilibrium exists between the inner and the outer pressure affecting the body tissues.

However, the pressure may limit the vertical range of motile forms. Although there is no harm done to eurybathic animals and they are not seriously affected. They are known to make daily vertical wanderings of up to 400 m, corresponding to pressure variations up to 40 atmospheres.

Harpooned whales are said to "sound" to a depth of 800 m, and the sperm whales must descend normally to great depths, since the large squids upon which they feed inhabit very deep water.

Water Movements

Most of the part is in constant motion with both regular and irregular pattern of flow. Keeping this in mind, the sea must be viewed as an environment that paves way for the same. The principal biological benefits derived from the circulation are:

- oxygenation of subsurface water;
- dispersal of wastes resulting from processes of metabolism;
- dispersal of plant nutrients and other variable elements essential to plant and animal growth; and
- dispersal of spores, eggs, larvae, and also many adults.

On the whole, there is a direct benefit from the circulation of water, yet there may be instances noted where some adverse conditions result either as incidental or as permanent features. The reason behind the incidental disturbances may be due to unseasonable shifts in the regular current system. For instance, it gives rise to the appearance of "El Nino" off the west coast of South America.

With reference to the above, the warm water of the Equatorial Countercurrent is carried southward along the coasts of Ecuador and Peru, which are normally bathed by cold currents. This results in a destruction of animal life along the coast, including many guano birds that depend upon the sea for food. The destruction of the marine birds is in huge numbers.

The inhabitants are carried by the moving water of the sea to the unfavorable living conditions. There are some permanent and some semi-permanent features of the current system that affects this. For example, the inhabitants of the gulf stream are ultimately perished as they are taken northward into the regions where the temperature of the water is lowered by the admixture of cold water or by the cooling in higher latitudes. The neritic forms contain larvae which gets dispersed into the offshore location uninhabitable to the adult.

The seawater is salty and there exists a surface current that strews the shores with defunct bodies of normally oceanic or offshore forms such as the coelenterate Vellela or the pelagic snail Janthina.

7.4.3. Extent of the Marine Environment

The part of the earth which is capable of sustaining life of both plant and animal, is known as the Biosphere. The biosphere is then subdivided into three principal divisions or habitats known as bio cycles. These include the terrestrial, the marine, and the fresh water biocycles. The biocycles having three different principle divisions has its characteristic types of ecological features and associations of plants and animals.

There are few animal species which at times can migrate from one place to another. This practice is commonly seen in salmon and the eel. The area of the ocean is about two and a half times the area of the land. Thus, it covers almost 71% of the earth's surface. However, not all the area of the ocean is worth living by the habitants of the sea. For this, the account has to be taken of the relative vertical range provided by the two main environments, the terrestrial and the marine.

The marine environment actually provides about three hundred times the inhabitable space provided by the terrestrial and the freshwater biocycles (Hesse, Allee, and Schmidt, 1937). However, the terrestrial environment provides space only in a shallow zone and that too at the immediate surface and to a depth of a few feet. There is space provided to some form of life from the surface even to the abyssal depth of several miles.

Although, there is a small fraction of the other two that the freshwater constitutes. The birds, insects taking entry into the aerial regions of the globe should be considered as a temporary journey. This is because the aerial region of the globe is not considered a separate biocycle.

7.5. WETLAND ECOSYSTEM

Wetland is the area that is the link between the water and the land. It comprises the areas that transition between terrestrial (land) areas and aquatic (water) areas. There is a rich variety of plants and animals found on the wetlands. The wetlands ecosystem represents a richly diverse web of flora and fauna interacting together.

Figure 7.4: Ecosystem of wetlands.

Source: Image by Needpix

Also, the wetland ecosystem is greatly influenced by the disturbance from outside. Since the wetlands are the most sensitive part of the ecosystem. These are mostly affected by the human intervention and the environmental damages done by the humans. There is a provision of natural storm barriers, environmental cleansers, and food and water resources for many forms of life by the wetland ecosystem.

The most valuable and productive ecosystem among the world is the wetlands. There is wide range of economic, social, environmental, and cultural benefits, in recent times classified as ecosystem services (Costanza et al. 1997).

The service includes from the maintenance and quality water supply, regulating atmospheric gases, sequestering carbon, protecting shorelines, sustaining unique indigenous biota, and providing cultural, recreational, and educational resources (Dise, 2009).

Although, the wetlands cover around only 1.5% of the Earth's surface, there is a disproportionate 40% of the global ecosystem services (Zedler and Kercher 2005) by the wetlands. They play a fundamental part in the local and the global water cycles. Wetlands are located at the heart of the connection between water, food, and energy. This is what is the challenge of the society in achieving the sustainable management.

There was a study conducted by the Economics of Ecosystems and Biodiversity for water and wetlands (TEEB) in 2013 to help the decision makers prioritize for the management and protection. The study translated the values of the ecosystem services into the dollar terms. For instance, the value of the inland wetland ecosystem was estimated at the US$44,000 per hectare per year.

The wetlands having higher value around the world were systematically drained and filled to support the agriculture. The other uses of the filled wetlands were for the urban expansion, and other developments. Unfortunately, there has been a loss of about 50% of world's original wetland area. This ranges from the relatively minor losses in boreal countries to extreme losses of >90% in parts of Europe (Mitsch and Gosselink 2000a).

The remaining wetlands in the developing and the developed regions of the world are under the increasing pressure of the direct and indirect human activities. The wetlands areas are continuously declining despite strong regulatory protection in many countries, (National Research Council 2001; TEEB 2013).

7.5.1. Importance of Protecting Wetland Ecology

The wetlands provide a balance between the species that live in wetlands and the environment around them. The wetland is affected by every aspect of the hydrology. The chemical and physical characteristics of the wetlands are decided by the flood, also the amount of oxygen existing in the wetlands. When this delicate balance unravels, wetlands, and their denizens suffer.

The wetlands of the world have suffered dramatic changes under the influence of human settlement, agriculture, and its runoff and the pollution emitted by the industries. The chemical balance of wetlands is the one on which the plants and animals survive.

Pollution disrupts the chemical balance of wetlands that plants and animals depend upon to survive. Wetlands provide flood control, storm barriers, clean water, and aquifer restoration. They also neutralize bacteria, absorb harmful chemicals and filter pollutants.

Wetlands provide foods such as rice, fish, cranberries, and other products with unparalleled economic importance. There is an estimate that at least 40% of the entire world's species resides in wetlands. Because without the healthy ecosystem of the wetland, many species on earth would suffer. Additionally, wetlands are known for its beauty and outdoor recreational activities for people to enjoy. There is however a need to find sustainable methods to conserve wetlands in an ever-changing world that remains paramount importance.

7.5.2. Types of Wetlands

The rooted vegetation of the wetlands is considered as emergent wetlands, and these include marshes and fens. The emergent wetlands include such plants as cattails, rushes, and water lilies. In scrub-shrub wetlands, the small saplings under 20 feet in height coincide with shrubs. The flooding in these areas may be seasonal or permanent.

Bog is one of the examples of a scrub-shrub wetland. It contains peat mats that float away from the shore. The water of bogs has higher acidity and lower oxygen levels. The bog water is not favorable for fish. The swamps get dominated by tall trees and their understory forested wetlands. Vernal pools represent temporary, shallow depressions formed from spring rains.

When the water table touches the surface, springs, and seeps occur. These seeps and springs provide another form of wetland which is important to plants and wildlife. Riparian wetlands are those areas that comprise alongside flowing waters such as streams and rivers. In these areas, soil erosion takes place easily.

1. Abiotic Factors of Wetlands

Abiotic factors in wetlands are the nonliving factors affecting wetlands ecosystems. The abiotic factors include water itself and its varying sources. The physiochemistry such as water and soil chemistry, hydrology or the effect of flooding and available oxygen are also considered as the abiotic factor.

Weather is another thing that comes under the abiotic factor that has a prominent effect on the wetland. Other than precipitation, the wetlands are affected through storms, winds, and the currents made by them in the larger water bodies adjacent to the wetlands.

In addition to precipitation, weather affects wetlands via storm winds and the currents made by them in larger bodies of water adjacent to wetlands. The other form of abiotic factor is the tide. Topography and the water level are the main factor that affects the wetlands adversely. There are numerous other abiotic factors including, alkalinity, sedimentation, erosion, turbidity (water clarity), nutrients, temperature, and physical dynamics such as ice scouring in colder climates.

Bedrock is also among the major abiotic factor of wetlands. This is the underling substrate's chemistry that directly affects the water and what kinds of species wetlands support. Wetlands have a direct adverse impact on wetlands. Another major abiotic factor is human intrusion via land use, agriculture, shipping, and urban development.

2. *Dominant Wildlife in Wetlands*

There is an extraordinary wildlife diversity found in the wetlands. Among the dominant wetland wildlife is the fishes and the crustaceans, migrating birds and waterfowl, and some mammal species such as foxes, minks, deer, and even bears. Wetlands are the spawning and nursery grounds for many fish. This is home to turtles, frogs, snakes, and other reptiles and amphibians.

Many of these animals are food for other animals and people. There are number of endangered and threatened wildlife species that resides in wetlands. Birds, mammals, fish or invertebrates are the dominant wildlife in wetlands. For survival, they rely upon the primary producers like aquatic vegetation. They ensure that the food webs remain intact both near and far from wetlands.

7.5.3. Why Are Wetlands Such Important Providers of Ecosystem Services?

The position of the wetland on the landscape enables the wetlands to provide high-value ecosystem services (Zedler 2006) as recipients, conduits, sources, and sinks of biotic and abiotic resources. These high-value services occur at the land-water interface. Mostly in the topographically low-lying positions that receive water, sediments, nutrients, and propagules washed in from upslope and catchment.

In the catchments, wetlands allow sediments and other materials to accumulate and settle, providing cleaner water for fish, wildlife, and people. The vegetation growth is promoted by the combination of abundant nutrients and shallow water in receiving wetlands. This enables the habitat and food for a wide range of fish, birds, and invertebrates.

There is a huge accumulation of floodwaters, retaining a portion, slowing flows, and reducing peak water levels. They have a cumulative effect on the flood abatement. The near-permanent wetness of wetland ecosystems is equally important.

There is a very low-level of oxygen in the saturated areas. This is seen particularly in the soil which is accessed by its roots and microorganisms (Sorrell and Gerbeaux 2004). The anoxic conditions promote changes in critical microbial processes resulting in anaerobic nutrient transformations. This results in the availability of the nitrogen for use by plants (nitrogen fixation). Then, the nitrates are converted into harmless gas, thereby improving water quality (denitrification).

The existence of anoxic and aerobic conditions in close proximity is a natural property of shallow water and wetlands (Zedler, 2006). The anoxic conditions also promote peat accumulation, locking up carbon. This in turn regulates atmospheric carbon levels and helps cool global climates (Frolking and Roulet, 2007).

7.6. CLASSIFICATION OF WETLANDS

Taxonomists has categorized organisms into species, genera, and higher taxa. The classification has been made according to the structure that facilitates the understanding of the relationship of the organism to each other.

The species of a flying bird can be identified by the ornithologists based on the characteristics of the field. The literature can be helpful in finding out about the species' life, culture, history, etc.

Ecologists create ecosystem classification units to represent repeating patterns of environmental conditions and the related biota in the landscape. These ecological units can be identified in the field. The information and interpretations, based on the experience of others, can be accessed once these have been defined.

The classification of wetlands enables the understanding of the ecosystem and helps in applying the principles of the ecosystem management. It also allows for the ordering, comparison, synthesis, mapping, and inventory of information and gives resource workers a common language to communicate results.

In British Columbia, Biological ecosystem classification (BEC) is the primary classification applied to forested sites. The environmental conditions or site potential is identified by the indicator species groups approach. A specific and detailed information on the plant species occurrence and the environmental conditions can be found out by the fundamental unit of bec.

There are some regions of the province for which there is wetland site associations. The site associations that have broadly similar ecological properties of hydrology, soils, and vegetation structure are described by the wetland class.

This level of the classification is useful for general discussion of wetlands (National Wetlands Working Group 1997). There are five wetland classes: Bogs, Fens, Marshes, Swamps, and Shallow Open Water. Bogs and Fens develop deep accumulations of peat and are referred to as Peatlands. However, swamps, and marshes occur occasionally on deposits of well-decomposed peat, but more commonly have mineral soils.

The wetlands are classified on the basis of regions, as follows:

7.6.1. Boreal Interior (Western BWBS, SBPS, SBS, SWB)

The central and sub-boreal interior consists of only 3% of the wetlands. Though there are areas with higher concentrations. The formation of peatlands in most wet depressions is promoted by the short, cool growing seasons.

Consequently, the subdued terrain may be covered by large expanses of fen ecosystems. In relation to the river and lake system, marshes, and swamps are common. Ecosystems of these areas have been described in Roberts (1984) and MacKenzie and Shaw (2000).

7.6.2. Hyper Oceanic Lowlands and Mountains (CWH, MH)

The windward side of the Coast Mountains, a wetland region experiences a hyper maritime climate characterized by extreme precipitation and mild summer and winter temperatures. In some areas, wetlands may cover up to 50% of the landscape. The geology is predominantly granitic and poor in minerals, with little or no glacial till.

Even though most sites receive groundwater seepage, wetlands are primarily bogs. These bogs are extensive on gently to steeply sloping terrain. These wetland areas have community types that is globally unique. Ecosystems of these areas have been described in Banner et al. (1988).

7.6.3. Temperate Coast (CDF, Dry CWH)

There is mild winters, warm summers, and moderate precipitation in the southeast corner of Vancouver Island, the Gulf Islands, and part of the adjacent mainland. The most common are the swamps and marshes in this landscape, though here are some extensive bog ecosystems that occurs (e.g., Burns Bog on the Fraser River delta).

In this landscape, the wetlands are not uncommon in the area (6% of the area). However, most of the wetlands have been modified by human activities, while some have been lost completely by the humans. Within this region, there is northern range limit of many plant and animal species.

This is combined with the mild climate, produces ecosystems not found elsewhere in the British Columbia. The ecosystem of Estuarine are distinct from those on the north coast. Ecosystems of these areas have been described in Ward et al. (1992) and Banner et al. (1988).

7.6.4. Taiga Plains (Eastern BWBS)

The taiga plains are extensive, poorly drained, glaciolacustrine deposits underlie the northeast corner of the province. The climatic conditions of the plains are boreal to subarctic with very cold winters and warm summers. The region has 30% aerial extent which is the highest concentration of wetlands in the province.

Bogs are found in great number of amount while fens and swamps occur along the sluggish streams that drain the region. These region faces the permafrost which is very common in the province. Ecosystems of these areas have been described in Zoltai et al. (1988).

7.6.5. Interior Rainforest (ICH, Wet/Very Wet Subzones of the SBS)

There is a blend of interior and coastal qualities in the interior belt of the wetland ecosystem. There is a great support in the wetland formation by the cool, wet climatic conditions. However, the wetland abundance has been restricted by the mountainous terrain. There are some large peatland or swamp complexes found in the valley bottoms. The ecosystem of these areas is described in the MacKenzie and Shaw (2000).

Figure 7.5: Interior rainforest.

Source: Image by Wikimedia Commons.

7.6.6. Semi-Arid Interior (BG, PP, IDF, MS)

The semi-arid interiors experience the high summer temperatures and low precipitation. There is peat formation in all the regions except at higher elevations. This is the reason marshes and swamps are the most common. Although, wetlands occur on less than 2% of the landscape, there is still some area in the Chilcotin Plateau which have higher wetland density.

During the summer season, evaporation may occur in the potholes and shallow lakes of the region. This results in the accumulation of salts and formation of distinctive ecosystems. The interior saline and alkaline sites are mostly restricted to this wetland region. The water resources are fewer and there is difference in the composition of the species and the structure from the adjacent upland habitats. Wetlands are particularly important in the semi-arid interior.

The Columbia marshes in this region has the largest wetlands which has high level of wildlife use. Additionally, there are certain wetland dependent wildlife species that are at higher risk such as the Great Basin spadefoot toad (Spea intermontana) and tiger salamander (Ambystoma tigrinum), which occur only in this region.

Most of the agricultural activities have altered the wetlands in this region. The activities such as crop production and grazing (Van Ryswyke et al. 1992) have adversely affected the wetland. Ecosystems of these areas have been described in Steen and Roberts (1988) and MacKenzie and Shaw (2000).

7.7. OTHER FACTORS AFFECTING THE ECOLOGICAL ROLE OF WETLANDS

Classification of the role of the ecology of the wetland serves to generalize the complexity of nature. However, there are additional factors that are not directly incorporated into the classification but are performing important ecological functions, such as providing fish and wildlife habitat, flood attenuation, or biogeochemical filtration. For example, the wildlife habitat quality of a wetland can be influenced by:

- Wetland class;
- Structural diversity and vegetation cover;
- Diversity of site associations;
- Presence of open water;
- Quality of adjacent upland habitat;
- Presence of adjacent wetlands;
- Regional scarcity of wetlands;
- Regional climate;
- Presence/absence of fish;
- Prey density; and
- Other unique habitat features.

Flood attenuation and filtering are a function of:

- Soil characteristics.

- Groundwater retention and flow rates, and
- Linkages within the watershed.

7.8. WETLAND INVENTORY AND ASSESSMENT

The vital components of the effective wetland management are inventory, assessment, and monitoring. The management decisions are supported by the essential data and information they provide (Dugan 1990, Finlayson, 1996a). Further, they assist in providing feedback on the management actions and in the implementation of principles and frameworks. This is to ensure that the right information is delivered to the managers and other decision-makers (Finlayson, 1996b).

There has been special attention given to the design and the implementation of the effective and integrated programs, as the inventory, assessment, and monitoring cannot be treated separately from management processes.

There is a global review of wetland inventories (GRoWI) for the Ramsar Convention. This has recommended future good practice and priorities (Finlayson and Davidson 1999; Finlayson et al., 1999). A framework has been developed by the Ramsar Convention in conjunction with the Environmental Research Institute of the Supervising Scientist, Australia for conducting wetland risk assessment as an integral component of management planning processes (van Dam et al., 1999).

For the monitoring purpose, the general principles and frameworks are developed. For example, in the Ramsar Convention (Finlayson 1996b) and the Mediterranean wetland initiative (MedWet) (Finlayson 1996c, Grillas 1996, Tomas Vives and Grillas 1996).

The workshop was held as a part of the 2nd International Conference on Wetlands and Development (Dakar, Senegal, 10–14 November 1998). These workshops were conducted to review past and current projects, and to develop recommendations for further implementation of wetland inventory, assessment, and monitoring.

The main objectives were to:

- By discussing the practical approaches, methodologies, and techniques, promoting the inventory, assessment, and monitoring of wetlands;

- Identify the working tools needed to improve delivery of wetland inventory and assessment; and
- Identification of priorities for wetland inventory and assessment in support of biodiversity conservation in Africa.

7.9. PREVIOUS CONFERENCE RECOMMENDATIONS

There are several major wetland conferences during the last 10 years that have produced recommendations for improved wetland inventory, assessment, and monitoring. The most relevant to the objectives of this workshop are summarized below:

7.9.1. Managing Waterfowl Populations (Matthews 1990) – IWRB, Astrakhan, Russia (Former USSR), 2–5 October 1989

This conference was convened by the International Waterfowl and Wetland Research Bureau (IWRB). The USSR State Committee for Environment Protection invited IWRB to lead the conference. There were six technical sessions held which discussed the distribution, status, and management, with an emphasis on hunting, of waterfowl.

This conference was the turning point in open contact between scientists from both sides of the "Iron Curtain" that divided Europe at that time. This conference was attended by more than 150 wetland scientists.

The recommendations specific to waterbird inventory, assessment, and monitoring are listed below:

- For better understanding of the population processes and as a basis for conservation actions, collecting the long-term data on waterbird populations;
- Regular feedback of data and information to be provided to the participants involved in surveys;
- Standardize and promote procedures used for data collection;
- Develop a manual for sampling waterbird populations and hold training workshops; and
- Address gaps in regional waterbird data sets with relevant national and local personnel.

7.9.2. Managing Mediterranean Wetlands and their birds for the Year 2000 and beyond (Finlayson et al., 1992) – IWRB, Grado, Italy, 3–10 February 1991

This conference was convened by the IWRB. The Region Friuli-Venezia Giulia, Italy invited IWRB to lead the conference. There were specific technical sessions and workshops that were used to develop a strategy to stop and reverse wetland loss and degradation in the Mediterranean (means a strategy to stop and reverse the wetland loss and degradation in the Mediterranean basin 1992).

Approximately 280 wetland experts from 28 countries attended the conference and principally addressed the wetland management issues. Recommendations specific to wetland inventory, assessment, and monitoring are listed below:

- Develop an inventory and assess all water resources of the Mediterranean; and
- Strengthen the monitoring procedures for Ramsar sites.

7.9.3. Old world and new world wetlands (Mitsch 1994) – Intecol Wetland Conference, Columbus, USA, 13–8 September 1992

The organizers of this conference were the School of Natural Resources, The Ohio State University, USA, on behalf of the International Society of Ecology (Intecol). Total of 905 wetland experts attended the discussion that included a variety of wetland topics as outlined in more than 500 papers.

The conference resulted in the adoption of several formal resolutions. Those relevant to wetland inventory, assessment, and monitoring are listed below:

- Develop and adopt an international classification system and guidelines for national wetland inventories.
- Establish and promote the use of an international committee to develop these (IWRB/IUCN/Ramsar).

7.9.4. Waterfowl and Wetland Conservation in the 1990s – A Global Perspective (Moser et al., 1993) – IWRB, St Petersburg Beach, Florida, USA, 12–19 November 1992

This conference was convened by the International Waterfowl and Wetlands Research Bureau (IWRB) at the invitation of the United States Department of the Interior, Fish, and Wildlife Service. For specific wetland and waterbird topics total six separate workshops were held. These workshops gave guidance to IWRB as it developed its work plan for 1993–1995.

The conference has approximately 240 wetland experts from 54 countries who attended the conference. The following recommendations specific to wetland inventory, assessment, and monitoring were made:

- Continue and extend the International Waterbird Census and collect information on waterbird habitats;
- The data collection with respect to the waterbird habitats must be continuous. This would help in coordinate international data and overviews;
- Disseminate the results of monitoring waterbird populations and wetlands;
- A set of criteria and a manual for measuring ecological change must be developed;
- For monitoring ecological change, preparation of training course material;
- Filling the gaps and reviewing the existence in national wetland inventories;
- For collecting the data and making the inventories more useful, development of techniques and manuals is recommended;
- Develop techniques for socio-economic assessments of wetland; and
- Develop a communications network for inventory and monitoring experts.

7.9.5. International conference on wetlands and development (Prentice & Jaensch 1997) – Wetlands International, Kuala Lumpur, Malaysia, 9–13 October 1995

The organizers of this conference was the Asian Wetland Bureau (AWB), the International Waterfowl and Wetland Research Bureau (IWRB), and Wetlands for the Americas (WA), and hosted by the Ministry of Science, Technology, and the Environment, the Selangor State Government and the Institute of Advanced Studies, University of Malaya.

Under this, four technical workshops were held to address global issues of wetland conservation and management in the changing world.

The participants count was around 290 delegates from 60 countries. Recommendations specific to wetland inventory, assessment, and monitoring are presented below:

- Establish links for open exchange of monitoring information between agencies and NGOs;
- For monitoring ecological change in wetlands, development of guidelines and standards;
- Means of evaluating and auditing wetland monitoring programs must be developed;
- Develop and maintain databases of monitoring programs in different regions;
- Designing of monitoring programs for providing technical support and training; and
- Standardized techniques and manuals for monitoring specific threats to wetlands are to be developed.

Common Recommendations From Previous Conferences

Recommendations from these meetings have been broadly consistent and can be grouped under six common themes:

- Long-term data collection on wetlands;
- The techniques, guidelines, and manuals to be standardized;
- Provision of training;
- Reviewing gaps and co-ordination of data collection;
- Developing and making greater use of networks; and
- The means to be developed for auditing existing effort.

The consistency of the recommendations may be partly attributable to a degree of consistency of attendance and participation of personnel from or associated with Wetlands International (formerly IWRB). Since the conferences were held in different geographic locations it is likely that many other, and different, interests were also represented, and the bulk of participants were not the same.

Furthermore, the meeting has broadened their focus from the water birds to the water birds and their habitats, and then to wetlands and their management as reflected in the title. The outcomes represent a decade of developing 5 interest in wetlands and an increasing awareness that wetland inventory, assessment, and monitoring was either needed or, where it existed, was inadequate.

7.10. WETLAND LOSS

Wetlands, marshes, fens, bogs, and swamps are the link between land and water. Wetlands consists of trees, grasses, shrubs, moss, and other plants. This is because these require at least some water coverage. There is abundance of essential ecosystem services in the wetlands, it includes:

- Water storage, storm protection, and flood mitigation;
- Water purification through retention of nutrients, sediments, and pollutants;
- Groundwater recharge;
- For many plants and animals, essential habitats including over 90% of the roughly 200 Great Lakes fish species that occur in the Great Lakes; and
- Shoreline stabilization and erosion control.

7.10.1. The Problem

The wetlands destruction is taking place at faster pace than in any other ecosystem. Wetlands are getting destroyed due to drainage, farming, housing, roads construction, and other development. Some of the lakes are lost due to the degradation of the wetlands. The most affected are the habitats of the wetlands who are only dependent on the wetlands for their survival. Many of the habitats lost their life due to the destruction in the wetland areas.

Figure 7.6: Preserving the wetlands is important for the survival of the ecosystem.

Source: Image by Nation Park Service.

There has been increase in the wetland destruction due to which there is flood and damage, nutrient runoff and water pollution, and shoreline erosion, and triggered a decline in wildlife populations. The wetlands destruction is detrimental to any region's economy: recreation like fishing, hunting, and wildlife.

7.11. ECOSYSTEM SERVICES

There are wide range of wetlands ecosystem service which is vital for human well-being. These services are classified as per the TEEB 2010. This relates to the benefits people obtain from ecosystems provisioning services. There is production of array of vegetation, animal, and mineral products that can be harvested for personal and commercial use.

However, the most significant of these is the fish, which is in demand from billions of people worldwide. These wetlands provide employment and income for at least 150 million people through a fishing industry (Ramsar 2009e). Another demanding food product is the rice that accounts for one-fifth of total global calorie consumption.

Other important food products grown in wetlands are sago and cooking oil (from palms from Africa), sugar, vinegar, alcohol, and fodder (from the Asian nipa palm), and honey (from mangroves). Wetland products also include fuelwood, animal fodder, horticultural peat, traditional medicines, fibers, dyes, and tannins.

7.12. BIOLOGICAL IMPACT ASSESSMENT SYSTEM

Threatening the ecology and putting human welfare at long-term risk. In 2009, the combustion of fossil fuels and production of cement sent more than 30 billion tons of carbon poured into the atmosphere. The combustion of coal itself representing 66% of the growth in emissions. About 51% of the carbon dioxide emissions only coming from China. Most of the energy-intensive manufacturing industrial goods are being imported by rich countries. It is necessary to conduct ecological impact assessment studies for the World Projects of power plants and cement plants and to design and develop greener energy systems. Global emissions of carbon dioxide from fossil fuel burning warm the earth, threatening the ecology and putting human welfare at long-term risk. In 2009, the combustion of fossil fuels and production of cement sent more than 30 billion tons of carbon poured into the atmosphere.

The combustion of coal itself representing 66% of the growth in emissions. About 51% of the carbon dioxide emissions only coming from China. Most of the energy-intensive manufacturing industrial goods are being imported by rich countries. It is necessary to conduct ecological impact assessment studies for the World Projects of power plants and cement plants and to design and develop greener energy systems.

Biological impact assessment is the process designed to systematically assess the potential or actual impact, including the hazards and benefits of the presence in the introduction of, or the entrance into a biological system of specific endemic or exotic organisms.

The emission of carbon dioxide from the fossil fuel globally warm the earth. It threatens the ecology and there is a huge risk in the human welfare. Earlier in the year 2009, the combustion of fossil fuels and production of cement sent more than 30 billion tons of carbon poured into the atmosphere. The combustion of coal represents 66% of the growth in emissions.

China is among one of the countries to produce more than 51% of the carbon dioxide. Due to the huge amount of carbon emission, there has been import of most of the energy intensive manufacturing industrial goods. There is a need to conduct ecological impact assessment studies for the World Projects of power plants and cement plants and to design and develop greener energy systems.

The main purpose of the environment impact assessment process is to encourage the planning and decision making while considering the ecosystem. This is done to arrive at the action which are economical and compatible. There are number of technical and professional considerations related to both the predictive aspects and the interpretation of the significance of anticipated changes. This is due to the prediction and assessment of the impacts on the biological environment.

There are many projects that has undesirable impacts on the terrestrial and the aquatic ecosystem. The adverse impacts could be habitat degradation through overgrazing practices; wetland drainage for agricultural, industrial or urban development projects; habitat loss. The fish of the ocean are also affected and the wildlife species due to the construction of hydropower projects.

The timber harvesting, recreational developments, and / or military training activities has resulted in the loss of critical habitat for endangered or threatened species. The emission of Sulfur-dioxide from coal-fired power plants has affected multiple aquatic and terrestrial ecosystems.

7.12.1. Important Elements of the Ecological and Biological Impact Assessment Process

The ecological impact assessment for prediction and assessment of impacts on the biological environment shall be undertaken to fulfill the following needs:

- The planning of the new projects, programs, and activities should take into account the environmental effects with respect to the ecological and biological aspects;
- There should be an environmental assessment that is needed to be carried out before commitments or irrevocable decisions are made for such of those projects which have an adverse effect on environment;
- The assessment results should be used in the planning and decision making and the implementation;
- To find the loss of wildlife and habitat;
- To find the disturbances of aquatic organisms and habitats;
- To find the destruction of vegetation; and
- To find the erosion and sedimentation.

The first step in the environmental impact assessment (EIA) process is to determine whether a project falls within the relevant act or regulations or not. Also, whether the development is likely to create significant environmental disruptions. The nature and magnitude of the environmental changes which is likely to occur and must be included in the EIS with respect to the biological class.

For example, vegetation, wildlife; sport and commercial fish species; and endangered species. With respect to endangered species, EIS includes census of flora and fauna in the affected area. The study is conducted on the life cycles, habitats, and food chains of representative species. This is with the particular emphasis on processes which control the behaviors of environmental components.

Then, the assessment is undertaken which leads to the preparation of the environmental impact statement (EIS). The EIS is open for public and it can be viewed at public hearings. EIA Statements describe the affected ecological setting without project.

7.13. CONCEPTUAL APPROACH

The conceptual approach is the study focused on ecological impact assessment.

Step 1: Identification of the biological quantity and quality impacts of proposed project activities.

Step 2: The existing biological environmental conditions are described.

Preparation of description of existing biological conditions and consideration of endangered or threatened species and critical habit.

Step 3: The relevant laws, regulations or criteria related to the impacts/ or conditions are procured; biological quality standards.

Step 4: The prediction of the impact of the biological environment is done.

Step 5: Assessment of biological impact significance.

Step 6: Identification and incorporation of mitigation measures.

7.14. CONSERVATION OF FLORA AND FAUNA

Flora is referred to as the plant life present in a particular region or habitat. Fauna is referred to as the animal life present in a particular region or habitat at a particular time. Biodiversity is a vast topic and it cannot be defined in one sentence.

Figure 7.7: Fauna on the earth.

Source: Image by Wikipedia.

Biodiversity is basically the life on the planet. The word biodiversity has arrived from the Latin word Bios meaning life and Diversitas meaning variety. Generally, it means the variety of life on earth. It includes genetic diversity, species diversity, and habitat diversity.

Figure 7.8: Flora on the earth.

Source: Image by Wikimedia Commons.

Diversity is the number of different items and their relative frequency. In terms of biological diversity, these things are organized at different levels. This range from the complete ecosystem to the chemical structure that are the molecular basis of heredity.

Thus, the term encompasses different ecosystems, species, genes, and their relative abundance." The area of flora, fauna, and biodiversity is quite interrelated. Flora and fauna form a major part of biodiversity.

Earth has variety of flora, fauna, and biodiversity. However, in the last few decades, there is a steady increase in the extinction of flora, fauna, etc. all over the world. The conservation of biological diversity is of paramount importance in the survival of the life on earth.

For preserving the continuity of the food chain on the earth, conservation of the biological diversity is necessary with that of conservation of essential ecological diversity. The genetic diversity of plants and animals has to be preserved for carrying forward the life on the earth.

It ensures the sustainable utilization of life support systems on earth. There is a vast knowledge of potential use to the scientific community. Also, there is a reservoir of wild animals and plants which is preserved. Thus, it enables them to be introduced in the surrounding areas if needed.

Biological diversity provides immediate benefits to the society. The benefits include recreation and tourism. Moreover, biodiversity conservation serves as an insurance policy for the future.

7.15. CONCLUSION

Ecosystem constitutes a region that has forest, grassland, desert, wetland or coastal area all around. In ecosystem, there is area that has specified and recognizable landscape which forms the ecosystem. There is wide variety of living forms on earth. Biological diversity is the variety of life forms on earth.

Often, biological diversity is known as biodiversity. It basically refers to the total number of different species on Earth. The collection of human being, animals, plants, etc., forms the biodiversity. The conservation of biodiversity is of utmost importance for the survival of the life on earth.

According to a database, there has been an estimate of the number of species on Earth that range from 5 to 7 million. However, it has been opposed by some scientists who estimate it as 30 million. The role of wetlands has been given importance in the chapter as most of the living organisms including humans depend directly or indirectly on the areas near the wetlands.

Similarly, the conservation of flora and fauna has its own importance. Earth has a variety of flora and fauna which has to be preserved for the future perspective. There is an increase in the rate of extinction in the flora and fauna which needs to be managed.

REFERENCES

1. Bhatia, C., (2011). *Conservation of Flora and Fauna*. [online] Conservation-of-flora-and-fauna.blogspot.com. Available at: http://conservation-of-flora-and-fauna.blogspot.com/2011/10/conservation-of-flora-fauna.html (accessed on 4 March 2020).
2. Clarkson, B., Ausseil, A., & Gerbeaux, P., (n.d.). *Wetland Ecosystem Services*. [online] Available at: https://www.researchgate.net/publication/260436894_Wetland_ecosystem_services (accessed on 4 March 2020).
3. Dotson, J. (n.d.). *Ecosystem of Wetlands*. [online] Sciencing. Available at: https://sciencing.com/ecosystem-wetlands-6164461.html (accessed on 4 March 2020).
4. Ecosystems. (2004). [eBook] Available at: https://www.ugc.ac.in/oldpdf/modelcurriculum/Chapter3.pdf (accessed on 4 March 2020).
5. FLOW. (n.d.). *Wetlands Destruction-FLOW*. [online] Available at: https://forloveofwater.org/issues/wetlands-destruction/ (accessed on 4 March 2020).
6. Mastorakis, N., & Iyer, V., (2012). [online] *Researchgate*. Available at: https://www.researchgate.net/publication/308696734_Ecological_Impact_Assessment_for_Prediction_and_Assessment_of_Impacts_on_the_Biological_Environment_with_respect_to_the_Indian_Scenerio (accessed on 4 March 2020).
7. The Ecology of Wetland Ecosystems, (2000). [eBook] Available at: https://www.for.gov.bc.ca/hfd/pubs/Docs/En/En45.pdf (accessed on 4 March 2020).
8. The Sea as a Biological Environment, (n.d.). [eBook] Available at: https://publishing.cdlib.org/ucpresseBooks/data/13030/6r/kt167nb66r/pdfs/kt167nb66r_ch08.pdf (accessed on 4 March 2020).
9. U.S., R., & Agarwal, N., (2015). *Biodiversity: Concept, Threats, and Conservation*. [online] researchgate. Available at: https://www.researchgate.net/publication/294876262_Biodiversity_Concept_Threats_and_Conservation (accessed on 4 March 2020).
10. Wetland inventory, assessment, and monitoring: Practical techniques and identification of major issues, (1998). [eBook] Available at: https://www.environment.gov.au/system/files/resources/68e7b642-509e-4811-a5e9-72ccd8e0b1b0/files/ssr161.pdf (accessed on 4 March 2020).

CHAPTER 8

Technology in Environment Impact Assessment

CONTENTS

In the chapter technology in environmental impact assessment (EIA), the role of various types of technologies in eradicating the impact of various activities on the environment is discussed. The chapter also discusses the technology in the treatment of wastewater. The chapter also explains the meaning of root zone wastewater treatment (RZWT).

The chapter also explains various techniques of soil pollution control management. The chapter also throws some light on the meaning of phytoremediation technologies. The chapter also discusses the meaning of bioremediation techniques. The chapter also discusses the role of remote sensing and GIS technology. Furthermore, the chapter also highlights the importance of noise control technology. In the end, the chapter explains the sustainable manufacturing process and its importance for the environment.

8.1. TECHNOLOGY IN TREATMENT OF WASTEWATER

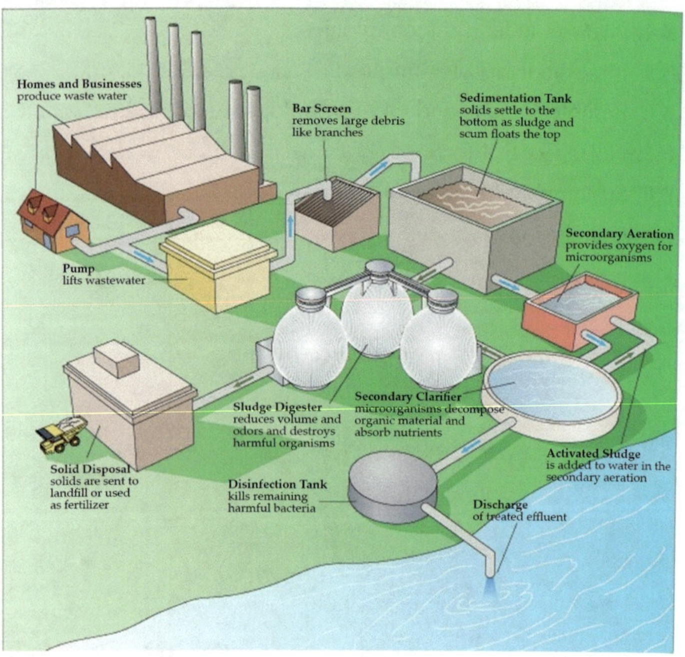

Figure 8.1: Wastewater treatment plant.

Source: Image by Wikimedia Commons.

Wastewater, which is the biggest waste stream from municipalities, pharmaceuticals, petrochemical, food, textile, polymer, agricultural, and paper industries and so on contain high contaminants of oil and salts of organic and inorganic compounds. This problem is considered as one of the major ecological problems that have significant impact on the environmental impacts when it is released into the ecosystem without any proper treatment.

Furthermore, with the rapid rate at which the industries are growing, the demand of freshwater supply has been dramatically increased and will continue to increase in the future. These industries are the ones that majorly contribute to depleting the quality of natural freshwater, although wastewater can be treated through various physical, biological, and chemical strategies. Unfortunately, the traditional available wastewater treatment methods are not capable enough to eliminate the contaminants entirely. Furthermore, the traditional wastewater treatment methods can also be expensive and thus not feasible to deal with a large amount of wastewater released by industries and the household sector. Therefore, the objective to ensure clean water and clean environment has forced many environmental protection agencies to set stringent discharge limits.

Conversely, there are always variations in wastewater qualities which resulted in having diverse impacts on the environment. Thus, having a proper wastewater treatment combined with primary, secondary, and advanced treatment strategies seems to more feasible or viable. The basic method concerned with treatment of wastewater is to sort the solids from the liquids via sedimentation or filtration process, whereas on the other hand, the secondary treatment eradicates the dissolved solids and other contaminants through biological process and chemical precipitation.

Then UV light or membranes play an important role in further treatment. After this, the treated water can be cost-effective for farmers and also cost favorable for the environment in a sustainable manner viz. irrigation, and agricultural purposes. In this section, evaluating the streamline flow of innovative wastewater treatment technologies for the reuse and subsequent sludge generation as an energy source is discussed. The biological treatment is the first and foremost method in the wastewater treatment process, followed by membrane technology and lastly the advanced oxidation process (AOP). All these methods are discussed as follows.

8.1.1. The Biological Wastewater Treatment Process

There are various complicacy and obstruction that comes in way of disposal of municipal solid wastes in the landfill sites as it carries the legislation with respect to proportion of biological waste that can be disposed in the landfill sites. The treatment process for these organic fractions is biological wastewater treatment.

These technologies increase the amount the waste that can be recycled and recovered from the total waste. The biological treatment is considered as one of the most important aspect in the wastewater treatment and is a widely used technique employed for industrial or municipal use for soluble organic components. Among all, the most widely employed method for sludge treatment is anaerobic digestion (AD).

8.1.2. Anaerobic Digestion (AD) Process

In this process, a large fraction of the organic matter (cells) is shattered into carbon dioxide (CO_2) and methane (CH_4), and this can be achieved without the presence of oxygen. About half proportion is then converted into gases, while the rest is dried and becomes a residual soil-like material. Angelidaki and Kougias informed that the end products of organic assimilation in the anaerobic treatment of waste are CO_2 and CH_4. The AD technology has been widely popular and used by many in the past few years with the applications of separately configured high rate treatment processes for industrial wastewater streams.

In the wastewater treatment settings, the AD technology has been employed in several instances throughout the world for biogas and bioremediation production. Biogas, one of the most sources of renewable energy, is produced with the help of the AD process, consisting largely of CH_4 and CO_2. As an alternative source of energy, the AD process produces biogas that can be primarily used as fuel in combined heat and power gas engines.

8.1.3. Membrane Technology

Membranes that serve as a basis of a thin layer barrier for size differential separation, are usually combined with biological or chemical treatment or with standalone systems in secondary treatment of wastewater settings. In a proper membrane process, there is generally a driving force, for instance, a semi-permeable barrier that acts as a controlling the rate of movement of components by rejection and fractional permeation through pores of different sizes.

The selective rejection and permeation are a function of the chemical affinity, and membrane pore size, which provides the support of a product stream devoid of target components. Due to the comparatively less energy requirement and wastewater treatability efficiency, there has been tremendous improvement in the membrane technology by the formation of new materials and configurations for industrial applications. Some of these applications include the removal of organic, microbial fuel cells, and inorganic components, pathogen removal, disinfection, and desalination.

8.1.4. Advanced Oxidation Process

Generally, there are two stages that are mainly used in the treatment of wastewater via a pre-treatment step involving physicochemical and mechanical systems to lessen the heterogeneous components of the effluents accompanied by a modern treatment process. The physiochemical process improved the efficiency of the advanced treatment by agglomerating the containments into a bigger size for easy removal or filtration.

However, degradation of emerging recalcitrant components with bioremediation and membrane in advanced treatment processes believed to be more complicated. So, in order to effectively solve this problem, AOP has been widely popular because of its ability to easily degrade a wide range of organic micro-pollutants. This process is concerned with the production of potent reactive hydroxyl radicals (E_0 = 2.8 eV) with photon energy and without additional chemical treatment.

Some of the examples are photochemical oxidation (UV/H_2O_2, Ultraviolet-UV/O_3), chemical oxidation (O_3, Fenton reagents), electrolysis, and sonolysis, heterogeneous photocatalysis (UV/TiO_2). These technologies usually make use of UV-A with long wavelengths of approximately 315–400 nm, and UV-C with short wavelength radiation of almost 100–280 nm for degradation of majority of environmental contaminants. Generally, UV/H_2O_2 and UV/O_3 processes consume significant quantity of oxidant, that make them unfeasible to operate.

On the other hand, the hazards that is linked with ozone (O_3) being an unstable gas restrict its application and is usually work with an ozone-water contacting device to transform the O_3 into its liquid form thus raising the overall cost of production.

However, the substantial operational conditions of ambient temperature and pressure and the use of a chemical stable and low-cost catalyst (TiO_2) are primarily attractive for entirely mineralization of contaminants and by-products. This process makes the heterogeneous photocatalysis techniques to be more valuable comparative with other ones. Some other benefits of using this technique include quick reaction rate, no sludge production, low cost and operating well at ambient temperature and pressure conditions.

8.2. WHAT IS A ROOT ZONE WASTEWATER TREATMENT?

It is generally seen that in countries where there is no proper management, sewage water finds their way into the rivers, making the entire river dirty. Many countries spend enormously for treating sewage water before it is getting released into the rivers. However, still there is need of more advance technologies to treat the sewage water in order to follow a minimum standard.

One of the key reasons that why many industries hesitate from employing a sewage treatment plant into their industry is because of huge investment required and cost and high maintenance of running the same. Therefore, most of the industries largely dependent on state-managed and operated STPs, which often breakdown because of the overload it bears.

While in some countries, large-scale conventional STPs are used, they usually fall into disuse because of their high functioning costs, equipment corrosion and lack of experienced operating staff. If sewage water is properly treated at each and every industry and residential colony, then majority of the problem can be easily alleviated.

Almost 70% of domestic water is released as wastewater, majority of which can be recovered and reused if it is treated properly. Hence, wastewater should be seen as a resource rather than as a waste. Wastewater can be primarily categorized into two types:

- Greywater: wastewater that is released from kitchens, bathrooms, laundry activities, etc.
- Blackwater: wastewater that is released from toilets containing fecal matter and urine, which is also called as "sewage"

Both grey water and black water can be easily treated with the help of PCARZWT and reused for non-potable applications such as kitchen gardening and toilet flushing.

Among the various types of STPs, PCARZWT is considered as the most innovative as well as eco-friendly. Some of the key benefits of using PCARZWT system is its low cost, no maintenance, and virtually zero energy.

8.2.1. What Exactly is Root Zone Wastewater Treatment?

'Root Zone' is a scientific term that includes all the biological activity among various types of microbes, water soil, the roots of plants, and the sun. It primarily entails planted filter-beds containing sand, gravel, and soil. The PCARZWT system is based on the application of nature's way of biologically processing domestic & industrial effluents.

This effective technology called Decentralized Wastewater Systems (DEWATS) was primarily developed in the year 1970s by Germany and since then has been extensively implemented and being used in different countries mainly in Europe and America.

The PCARZWT system is based on the application of biological and physical-treatment processes to remove pollutants from wastewater. Because of its natural way of eradicating the pollutants from wastewater, there is no need to add any input such as mechanical pumps, chemicals, or external energy. It helps in saving both the energy as well as maintenance costs.

To accomplish this, the following steps are undertaken by the PCARZWT system:

- **Pre-treatment done in a Settler**: a device which is used to separate the liquid from the solid.
- The first treatment takes place in an **Anaerobic Baffled Reactor**: a device that is equipped with various identical chambers through which the effluent surpasses from top to bottom.
- The second treatment happens in an **Anaerobic Filter**: a device that is integrated with a filter material (cinder), through which the effluent surpasses from top to bottom.
- The third treatment takes occurred in a **Planted Gravel Filter**: a system that is integrated with gravel material and planted with water-resistant reed plants, that performs the role of providing oxygen to the passing effluent.

The PCARZWT system basically take into notice the natural slope of the ground, so that water easily moves from one unit to another without any external forcer required such as motor pump.

Once the reed plants create an established stand, generally after the first growing season, the reed bed does not require any maintenance. The plant foliage will then naturally blend into the landscape, ever-changing with the seasons and creating a pleasing sight as well.

Table 8.1: Difference between Conventional Treatment Plants and Root Zone Treatment Plants

Conventional Water Treatment Plant	Root Zone Treatment Plant
A lot of sewage aeration tanks powered with mechanical energy do not work because of power cuts, technical problems, and poor maintenance	By its treatment and construction features DEWATS is designed to limit operation and maintenance requirements and in the same to ensure high treatment standards even for shock loads
It requires a lot of external energy such as motor pump	It uses very minimal external energy
There is need of pumps in this system	It works on the application of gravity, avoiding the use of any pumps, as well as valves
Regular or day to day maintenance is required	There is no maintenance required
It carries high operating costs	It fetches low operating costs

Source: Table from ecodeaz.

8.3. SOIL POLLUTION CONTROL MANAGEMENT TECHNIQUES

Figure 8.2:An example of soil erosion.

Source: Image by Wikipedia.

Soil is one of the most valuable important resources of the nature. Life and living on the earth would be impossible if there is no healthy soil available on the earth. The primary role of soil is to provide food. Almost 95% of the food is derived from earth and soil play an important role in it.

Engaging in decision making, planning or strategies for having healthy and productive soil is essential for human survival. Entrance of materials, energy, essential nutrients or biological organisms into the soil will cause changes in quality of soil. This problem causes soil to remove from its natural state.

The soil is composed of two parts. 1-Soil Living Part, 2-Soil Dead Part.

The dead part of the soil mainly comprises of minerals and weathered rocks which can be attained from the decay of animals and plants that is called humus or organic matter, and air and water are considered in this part. But the live soil refers to the soil that mainly consists of small animals such as worms and insects, bacteria, fungi, and other microbes which are grown in the live soil.

The soil is basically the mixture of 50% of organic and inorganic matters, and 50% of water and air, which fills existing vacant spaces of the soil and keeps live organisms of the soil.

Soil Pollution: The soil can be polluted because of the number of the activities of humans which may have direct impact on the quality of soil. Most of these pollutions are caused by vehicle accident which moves contaminants.

The other pollutants that are responsible for soil pollution include trucks, cars, and airplanes that do not move the waste; rather, they majorly comprise materials such as fuel that could be responsible for causing soil pollution as a result of pouring and emitting them from the vehicle.

Some other factors such as human activities are largely responsible for polluting the soil. Dumping of toxic substances, for instance, various categories of solvents, detergents, and colored materials will extend earth and soil pollution.

8.3.1. Role of Soil Purification

Soils are believed to be purifications of the nature. Moreover, apart from the primary supplier of food, soil also possess purification property. This soil property can be attained as a result of their physical properties (water infiltration operation from pores), their chemical properties (evaporation and surface absorption) and their biological properties (corruption of organic matters and decomposition).

There are various types of chemical and physical processes that are responsible for formation of different types of soil. During the chemical processes, organic, and inorganic matters experience decomposition stage and thus transform. Meanwhile, physical processes produce the moved materials and different horizons of the soil. Soil is the origin of the universe and also human heritage for the future and posterity.

8.3.2. Soil Pollution Control Techniques

Soils play a very important role for providing food to large number of people. The good quality of soil is of utmost importance to ensure the continuous and consistent supply of food. Apart from providing food, it also plays the role of purification of the nature. It is the duly responsibility of those to maintain the quality of soil that are directly associated with soil. It can be done with the help of various soil pollution control techniques. Some of the soil pollution control techniques are discussed as follows:

Controlling Oil Pollutions in Soil

Oil materials and their derivatives result in causing soil pollution because of transport or storage. In case higher amount of oil materials will somehow enter into in-depth of soil, it will be more difficult to remove the pollution from soil. Some bacteria and microorganisms in soil can cause decomposition of oil materials.

Hereunder are regarded as the ways to control effects of oil pollution:

1. Put every effort to prevent rapid spread of oil;
2. Improve the ventilation in soil with the help of mixing and plowing;
3. Providing the essential food nutrients to the soil such as phosphor and nitrogen; and
4. Combining soil with microorganisms that help in decomposing of oil materials.

Controlling Pollutions Caused By Waste in Soil

Waste is considered as one of the most vital sources of soil contamination. Waste can penetrate into the ground and thus pollute the essential water resources as well.

Methods of waste disposal include: dumping, incinerating, and recycling.

In dumping method, some areas are specially allocated for the disposal waste and are called as "Land field." In this method, waste is dumped below the ground surface, with the objective to not be seen from ground surface, however, it creates a number of subsequent problems.

Some of the problem that arise because of this activity include pollution of water resources, releasing of bad odor into and poisonous methane gas fresh air which increase likelihood of fire danger, creation of harmful insects and organisms.

In order to have a control on a soil pollution caused primarily by the waste, the following techniques can be implemented:

- Application of effective technology for dumping waste such as compressing and filling of holes and other openings;
- Dumping waste at a level which is higher than the underground water levels;
- Creating impenetrable layers in building of land fields;
- Creating drainage system that helps in collection of leachates; and
- Using the gases produced in land fields.

In incineration method, all wastes are accumulated from its original source of generation and then they are put on fire away from the residential area. Incineration method is considered as one of the worst methods of waste disposal, because, incineration releases high amount of poisonous gases which when entered into air pose significant threat to the environment and also to the people living nearby. After incineration, ash of waste is remained and will create visual pollution.

Recycling is believed to be the best method of waste disposal among all till present date. With storing some waste and reusing them, human can significantly reduce the amount of waste and energy that is being used for the creation of new material. In this method, not only the problem of more waste is prevented, but more cost will also be saved remarkably.

Controlling Pollution Caused By Industrial Activities in Soil

This method refers to all the pollutant that come in contact of soil by the factories. Some of the waste that include in this are discussed as follows: Wastes produced by power plants and steel industries, wastes of steel manufacturing industries, wastes of chemical industries, wastes of oil industries (refining and extraction), wastes of metalworking industries, wastes of cellulose, wood, and paper manufacturing industries, waste of food industries as well as wastes of leather production industries.

The gathering of heavy metals in soil is also one of the major problems that is caused by industrial pollutions. These metals basically comprise of cadmium, lead, silver, and mercury that have significant harmful impact on the living organisms and have repeatedly caused environmental disasters.

Some of these effects are stated as follows:

- Disturbance of biological activities of soil;
- Toxic effects on plants; and
- Harmful effects on human being as a result of entrance of materials to the food chain.

There are three primary methods for soil decontamination from industrial wastes as follows:

- Soil can be excavated up to a particular depth and the excavated soil can be occupied from the region and then, it can be restored.
- The restoration of soil is possible at the same area.
- Keeping soil in the area is the other method. Under such circumstances, it is suggestable to add auxiliaries to the soil in order to prevent spread of infection to animals, the plants, as well as human.

Usually, a large plastic is drawn on the soil with the objective to prevent spread of soil pollution, to prevent disseminate of pollution to the other regions and to prevent water from penetrating into the soil.

Some of the Feasible Soil Restoration Methods Are

Make use of water to eradicate pollutants from the soil, using aerial and chemical solvents, removing pollutants with incineration, helping natural organisms for breaking down atoms of pollutants, adding materials to the soil in order to prevent the pollution being spread to the other regions.

Role of Plants in Controlling and Decreasing Soil Pollution

Pollution that are the result of exhaust of cars, industries, homes, and departments' heating devices and appliances will pose serious threat to the health of human beings, water, animals, soil, and air.

Some of solid evidence of the said claim are the recent increase in disease among human beings, eradication of many plants and endanger of many species of animals.

Because of this, environmentalists, and biological scientists are forced to think of natural ways (biologic) for finding solution to the problem of pollutions. The use of plants is the best feasible biological way, because, using plants is a non-technical and cost-effective method in terms of technology.

8.4. PHYTOREMEDIATION TECHNOLOGIES

Phytoremediation refers to the process in which plants are used in a way to clean or purify contaminated environments. Plants can play active role in cleaning up of many types of contaminants including pesticides, metals, explosives, and oil.

However, they work with their full efficacy at a place where contaminant levels are low because high concentrations may restrict growth of plant and take too long to clean up. Plants also help prevent rain, wind, groundwater flow from disseminating contaminants to other areas such as surroundings or to deeper underground.

8.4.1. How Does It Work?

- Certain plants are able to break or remove down harmful chemicals from the ground when their roots take in nutrients and water from the contaminated soil, groundwater or sediment. Plants can able to clean up or purify the soil from contaminants as far as their roots can reach using natural processes to:
- Store the contaminants in the stems, roots, or leaves.
- Transform them into less harmful chemicals within the root zone or within the plant level.
- Convert them to vapors, that can then release into the air.

- Sorb (stick) contaminants onto their roots where very minor organisms called "microbes" (like bacteria) that exists in the soil break down the sorbed contaminants to less harmful ones.

8.4.2. How Long Will It Take?

The Phytoremediation process may take many years to clean up a site. There are various factors on which the cleanup time depends. For example, the process of phytoremediation will take longer time where:

- Concentrations of contaminant are relatively high;
- The contaminated area is either deep or large;
- Plants that take longer than usual time to grow are used; and
- The growing season is of shorter duration.

All these factors vary greatly from site to site. Plants may have to be replaced in case they have been damaged by extreme weather conditions, pests, or animals. This will also result in adding time of cleanup.

8.4.3. Why Use Phytoremediation?

It is generally seen that environmental protection agencies uses phytoremediation for a number of reasons. First of all, it carries natural plant processes and requires less labor and minimal equipment comparative with other methods as plants do most of the work.

Also, the site can be easily cleaned without the much need of digging up and hauling soil as well as pumping groundwater, thus saving huge amount of energy. Trees and smaller plants that are used in the phytoremediation process help in controlling soil erosion, make a site more lucrative, improve air quality of nearby regions, and reduce the overall noise level.

8.5. BIOREMEDIATION TECHNIQUES

Within the past few years, there has been significant development in bioremediation techniques with the primary objective to effectively restore polluted environments in an eco-friendly way, and at a very low cost. Researchers have developed various types of bioremediation techniques; however, due to complexity in nature and/or type of pollutant, there is no single bioremediation technique that act as a single package to restore polluted environments.

Autochthonous (indigenous) microorganisms being present in polluted environments play very crucial role in addressing majority of challenges associated with biodegradation and bioremediation of polluting substances. According to Verma and Jaiswal 2016, the present environmental conditions are suitable for their growth and metabolism.

Some of the key benefits of using the bioremediation technique is that it is cost-effective and also side by side environmentally friendly compared with both chemical and physical methods of remediation. Thus far, various good definitions have been given to bioremediation, with a major focus on one of the processes (degradation).

Nevertheless, in most of the cases, the term biodegradation can be used interchangeably with bioremediation; the former is a term, which applies to a process under the latter. In this review, bioremediation can be defined as a process, that primarily dependent on biological mechanisms to reduce (detoxify, degrade, transform or mineralize) concentration of pollutants to an innocuous state. The approach of removal of pollutants depends largely on the nature of the pollutant, which may include chlorinated compounds, agrochemicals, dyes, heavy metals, greenhouse gases, hydrocarbons, plastics, nuclear waste, and sewage.

8.5.1. Ex Situ Bioremediation Techniques

These techniques are majorly concerned with excavating pollutants from the pollution affected sites, and afterwards shift them to another unit for further treatment. There are number of factors on which Ex situ bioremediation techniques are based on: depth of pollution, the cost of treatment, type of pollutant, geographical location, degree of pollution, and geology of the polluted site. Performance criteria, that also serve as a basis for ex-situ bioremediation techniques have been described (Philp and Atlas 2005).

8.5.2. Biopile

Biopile-mediated bioremediation performs a key role in above-ground piling of excavated polluted soil, accompanied by nutrient amendment, and occasionally aeration to the overall bioremediation by surging microbial activities.

Some of the components of this technique are irrigation, aeration, nutrient, and leachate collection systems, and a treatment bed. In the past few years, the use of this ex-situ technique has been considerable increased because of its unique features which includes cost effectiveness, that ensures effective biodegradation on the condition that temperature, nutrient, and aeration are sufficiently controlled (Whelan et al. 2015).

The application of biopile to polluted sites also play crucial role in restricting volatilization of low molecular weight (LMW) pollutants; it can also be effectively used in remediating polluted extreme environments such as the very cold regions (Dias et al. 2015; Gomez and Sartaj 2014; Whelan et al. 2015).

8.5.3. Windrows

As one of ex situ bioremediation techniques, windrows primarily depend on periodic turning of piled polluted soil in order to improve overall bioremediation by enhancing the degradation activities of transient and/or indigenous hydrocarbon clastic bacteria present in polluted soil.

The periodic turning of polluted soil, along with addition of water bring about increase in even distribution of pollutants, aeration, microbial degradative activities, and nutrients and thus triggering the rate of bioremediation, that can be achieved through biotransformation, assimilation, and mineralization (Barr 2002).

Windrow treatment in comparison with biopile treatment, suggest higher rate of removal of hydrocarbon; however, the overall efficiency of the windrow towards removal of hydrocarbon removal majorly depend on the type of soil, which was reported to be more friable (Coulon et al. 2010). Nevertheless, because of the periodic turning concerned with windrow treatment, it is not considered as the viable option to adopt in remediating soil polluted with toxic volatiles.

The use of window treatment has been implicated in CH_4 (greenhouse gas) that discharged at a time of development of anaerobic zone within piled polluted soil, which generally happens following reduced aeration (Hobson et al. 2005).

8.6. REMOTE SENSING AND GIS TECHNOLOGY

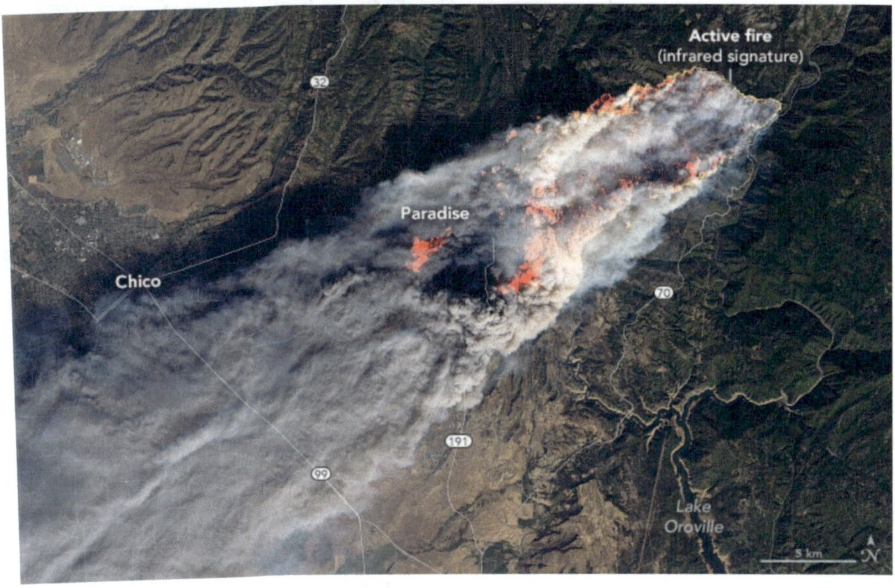

Figure 8.3: Assessing conditions of the earth through remote sensing and GIS technology.

Source: Image by NASA Earth Observatory.

Nowadays the field of Remote Sensing and GIS has become widely popular and recognized by many with rapidly expanding opportunities. Many organizations that are interested in this field are investing heavily to gain the advantage from the same.

Here the question arises why these fields have gained momentum in recent years. Two major reasons for this is that:

- Nowadays researchers, scientists, even common people are showing significant interest in order to gain better insight about the environment. The term environment means geographic space of their study area and the events that take place there.

In other words, it is now realized that geographic space along with the data describing it, is part of our everyday world; almost every decision taken by scientists or researcher with respect to environment is influenced or dictated by some fact of geography.

- The recent advancement in sophisticated space technology (that provide ample data and information about spatial data), along with reduction in cost of computer software and hardware (which can manage these large amounts of data) has made Remote Sensing and G.I.S. economical to not only complex environmental/spatial situation but also affordable to large group of people.

8.6.1. Remote Sensing

Literally Remote Sensing can be defined as the process of attaining information about an area, object, or phenomenon without directly coming in contact of it. Remote Sensing is a wider term, there are number of things that come under Remote Sensor, e.g., Fathometer, and Seismographs, etc.

Without coming in direct contact with the focus of earthquake, seismograph can successfully estimate the intensity of earthquake. In a similar way, without coming in contact with the ocean floor, fathometer help in measuring its depth. However, modern Remote Sensing play the role of obtaining information about earth's land and water surfaces with the help of reflected or emitted electromagnetic energy.

In other words, remote sensing can be defined as a process of detecting, monitoring, and assessing the physical characteristics of an area by computing its reflected and emitted radiation at a particular distance (majorly from aircraft or satellite). For this purpose, advance cameras are used that collect remotely sensed images that help researchers "sense" things about the Earth. Some examples of remote sensing technology to get better idea about this are discussed as follows:

- Cameras embedded on airplanes or on satellites click perfect images of wide areas on the Earth's surface, thus making it possible to see as wide as one can while standing at a same position on the ground.
- Sonar systems on ships can be used to map the area of the ocean floor without the much need to travel to deep inside the ocean.
- Cameras that are embedded on satellites help in taking images of temperature changes in the oceans.
- Some specific uses of remotely sensed images of the Earth include:
- With the help of remote sensing technology, it now become possible to map the large forest fires from space, allowing rangers to gain an insight on a much larger area than from the ground.

- Assessing the conditions of clouds to help ascertain the weather or watching erupting volcanoes, and help watching for dust storms.
- Keep a track on the growth of a city and transformation in farmland or forests over several years or even decades.
- Finding and mapping of the rugged topography of the ocean floor (e.g., deep canyons, huge mountain ranges, and the "magnetic striping" on the ocean floor).

8.7. NOISE CONTROL TECHNOLOGY

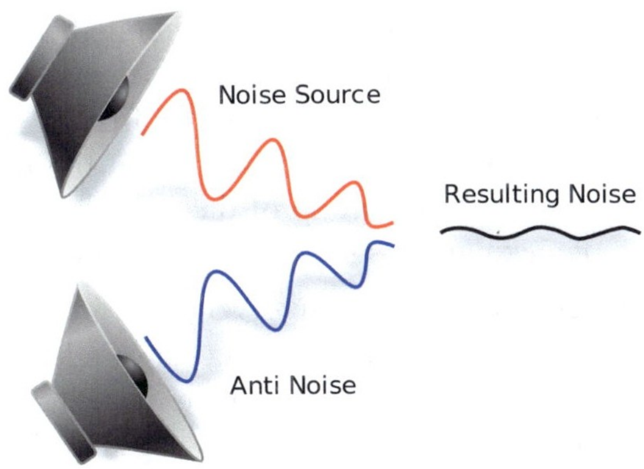

Figure 8.4: Active noise reduction.

Source: Image by Wikimedia Commons.

Noise control technology as the name suggests aimed at the controlling or reducing the noise at acceptable levels by action on the work environment. Such action includes all the measures strategies, and plans implemented with the objective to reduce noise being generated, and/or will reduce the noise transmission through the air or through the structure of the workplace.

Such measures include modifications or advancement in machinery, the workplace operations or activities, and the layout of the workroom. In fact, one of the effective and best way for noise hazard control in the work environment, is to eradicate it from roots, for instance, from the source of its generation, either by direct action on the source or by its confinement.

Practical considerations should be given equal importance; it is not suggestable to directly implement a global control program all at once. The most sensitive problem needs to be given first priority. In some instance, the solution may be found in a combination of measures which by themselves would not be viable; for example, to attain a desired reduction through environmental measures and to complement them with personal measures (for instance wearing hearing protection for only 2–3 hours), bearing in mind that it is extremely difficult to make sure that hearing protection is properly fitted and properly worn.

8.7.1. Control of Noise at the Source

To fully understand noise control, it is recommended to have some basic knowledge about some fundamental concepts. To restrict noise from its source of generation, it is foremost important to, first of all, determine the root cause of the noise and secondly to think about the measures that can be taken to limit the noise. Modification of the energy source to reduce the noise generated often considered as the best way to control the noise.

For example, where impacts are involved, like in punch presses, any lessening of the peak impact force (even at a cost of a longer time duration over which the force acts) will dramatically reduce the noise generated. Generally, when a choice of mechanical processes is probable to achieve a given task, the best choice, from the perception of minimal noise, will be the process which reduces the time rate of change of jerk or force (time rate of change of acceleration).

Control of noise at the source of its generation may be done either indirectly, i.e., generally, or directly, modification in the design of machinery to reduce the sound level.

8.7.2. General Source Noise Control Can Involve:

- **Maintenance:**

a. Adjustment or adjustment of worn or loose parts.

b. balancing uncertain or inconsistent equipment.

c. time to time providing lubricants to the moving parts.

d. use of cutting tools that are properly sharpened or shaped.

- **Substitution of materials** It is considered as the best possible solution which result in reducing the overall noise level (e.g., plastic for metal), a good example seems to be the viable replacement of steel sprockets in chain drives with sprockets that are made from flexible polyamide plastics.

- **Substitution of equipment:**

a. electric for pneumatic (for instance, hand tools).

b. use of stepped dies instead of single operation dies.

c. rotating shears should be the first choice than square shears.

d. hydraulic instead of mechanical presses.

e. presses instead of hammers.

f. use of belt conveyors instead of roller conveyors.

- **Specification of quiet equipment.**
- **Substitution of parts of equipment**:

a. updating the gear teeth, by substitution of spur gears with helical gears -resulting in almost 10 dB decline in noise level.

b. substituting straight-edged cutters with spiral cutters (for instance woodworking machines, which result in almost 10 dB(A) reduction in noise level).

c. substituting the gear drives with belt drives.

d. replacement of metal gears with plastic gears as they are more soundproof (beware of additional maintenance problems).

e. replace solid or steel wheels with pneumatic tires.

- **Change in methods of work operations**

a. in building demolition, substitute the use of ball machine with selective demolition.

b. replace pneumatic tools by modifying the manufacturing criteria, such as doing holes in concrete rather instead of cutting after the formation of the concrete component.

c. always use the remote control of equipment that are more noise centric like pneumatic tools.

d. separate noisy workers in time but try to put all the noisy operations in a particular area, far away from non-noisy processes.

e. set the speed of the machine at low or medium-level so that it does not produce more sound.

f. try to prefer large, slow machines instead of smaller ones with faster speed.

g. try to shorten the width of tools in contact with workpiece, it is expected that there will be almost (2 dB(A) reduction for each halving of tool width.

- **Substitution of processes.**

a. trying to use mechanical ejectors in place of pneumatic ejectors.

b. hot for cold working.

c. pressing in place of forging or rolling.

8.7.3. Source Control By Design Involves

- **Reduction of mechanical shock between parts by:**

a. Updating or altering the parts in order to prevent ringing or rattle.

b. Trying to use a flexible height collector for parts falling into a bin, so that there can be reduction in the impact speed and thus resulted in reduction of noise level.

c. Trying to use a flexible height of conveyor and rubber flaps in order to restrict the fall height of the parts.

d. Lining of tumbling barrels, metal chutes, parts collecting bins, hoppers, etc. with elastic material, e.g., hard rubber, cork, the conveyor belt material, plastic, with the type of material that should be as soft as possible but sufficiently hard to sustain the specific operating environment without wearing out prematurely. In extreme cases, another viable option is to line the chute or bin with a thin layer of viscoelastic material such as silastic, or silicone rubber, acting as a sandwich between the bin and a second layer of steel or other abrasion resistant material, with the latter layer having equal thickness to the wall of the bin or chute;

e. Covering metal wheels, metal tables, etc. with a flexible material like rubber.

f. Make use of conveyor belts rather than chutes in order to prevent any noisy falls.

- **Reducing the level of noise that may be because of out-of-balance by:**

a. trying to create a balance between moving parts.

b. use of vibration dampers and absorbers tuned to equipment resonances.

- **Lessening the amount of noise that caused by friction between metal parts by:**

a. always try to use lubrication or soft elastic interspacing (the perfect example of a noisy door which can be fixed by applying oil, which also increase the efficiency level).

8.8. SUSTAINABLE MANUFACTURING PROCESS

Sustainable manufacturing can be defined as the approach of creating manufactured products through the process which minimize the overall negative environmental impacts, conserve natural resources and energy and are also safe for employees, consumers, and communities and are economically sound. Traditionally, manufacturing can be defined as the sum of process that transform the raw material into finished good which is in final form that is ready for use by the ultimate consumer.

Sustainable production majorly focuses on a life cycle perspective in the manufacture, reusing, recycling, and disposal of goods and services, rather than the traditional emphasis on discrete activities. It always favors continuous and large-scale production in order to improve the overall efficiency, thus saving both the energy as well as resources.

Therefore, sustainable production can be defined as the formation of goods and services using systems and processes that are conserving energy, non-polluting, and natural resources. Nowadays, sustainable manufacturing has emerged beyond the life cycle view and the businesses or organizations that have successfully implemented sustainable manufacturing has been significantly benefitted from the same. Some of the benefits include business excellence, financial performance, and relationship with employees, clients, and other stakeholders:

1. **Financial performance:**

a. Increase the overall sale volume.

b. Improve productivity and efficiency by reducing use of resource, thus minimizing waste.

c. Reduce expensive and dependence or hazardous materials.

2. **Business excellence:**

a. Always try to follow up the regulations and be aware if there is any alteration or modification in regulation.

b. Win access to capital.

c. Implement smart strategies and planning.

3. Relationship with stakeholders:

a. Enhance repetition that help in building more strong relationship.

b. Demonstrating green know-how and setting a positive example.

c. Improve morale of employees which help in improving the absenteeism and retention rate.

d. Always try to build better community relations.

It is generally seen that companies all over the world are facing the problem of increase in cost of materials, energy, and compliance coupled with higher expectations of investors, customers, and local communities because they throw away valuable natural resources as waste, so in return they expect the same from business. Waste disposal often have negative impact on the local air pollution as well as greenhouse gas emissions. Sustainable waste management is essential because of the following reasons:

- It helps in saving of valuable natural resources, thus preserving them for the future generation.
- It helps in avoiding unnecessary emission of gas which further helps in slowing down the rate at which environment is degrading.
- It helps in ensuring the health of public as well as protecting the natural ecosystems.

8.9. CONCLUSION

In the end, it is concluded that technology play very important role in eradicating or slowing down the impact of human activities on environment. With the industrialization and increase in expectation of customers, companies have been extracting natural resources from environment at a faster rate. It results in depletion of natural resources, thus leaving less resources for future generation. There are various types of technology and each play different role in providing support to the environment. It depends on the business or organization which type of technology they need to use too get the maximum benefits in terms of cost, return on investment, energy as well as time.

REFERENCES

1. A review of phytoremediation technology: Heavy metals uptake by plants, (2018). [eBook] Available at: https://iopscience.iop.org/article/10.1088/1755-1315/142/1/012023/pdf (accessed on 4 March 2020).
2. Anon, (2019). *Treatment of Water and Wastewater for Reuse and Energy Generation-Emerging Technologies*. [online] Available at: https://www.intechopen.com/books/water-and-wastewater-treatment/treatment-of-water-and-wastewater-for-reuse-and-energy-generation-emerging-technologies (accessed on 4 March 2020).
3. Conserve Energy Future, (n.d.). *10 Magnificent Ways for Prevention and Control of Soil Pollution-Conserve Energy Future*. [online] Available at: https://www.conserve-energy-future.com/soil-pollution-prevention-control.php (accessed on 4 March 2020).
4. EcoIdeaz, (n.d.). *What is a Root Zone Waste Water Treatment?* [online] Available at: https://www.ecoideaz.com/innovative-green-ideas/whats-a-root-zone-waste-water-treatment (accessed on 4 March 2020).
5. Engineering Noise Control, (n.d.). [eBook] Available at: https://www.who.int/occupational_health/publications/noise10.pdf (accessed on 4 March 2020).
6. GRACE Bioremediation Technologies' DARAMEND'. Bioremediation Technology, (n.d.). [eBook] Available at: https://clu-in.org/download/toolkit/540r95536a.pdf (accessed on 4 March 2020).
7. Natural Technologies of Wastewater Treatment, (2014). [eBook] Available at: https://www.gwp.org/globalassets/global/gwp-cee_files/regional/natural-treatment.pdf (accessed on 4 March 2020).
8. ncbi.nlm.nih.gov. (2020). *Bioremediation Techniques-Classification Based on Site of Application: Principles, Advantages, Limitations, and Prospects*. [online] Available at: https://www.ncbi.nlm.nih.gov/pmc/articles/PMC5026719/ (accessed on 4 March 2020).
9. Phytoremediation, (1996). [eBook] Available at: https://clu-in.org/download/toolkit/phyto_o.pdf (accessed on 4 March 2020).
10. Remote Sensing and Geographical Information System (GIS), (n.d.). [eBook] Available at: http://ags.geography.du.ac.in/Study%20Materials_files/Punyatoya%20Patra_AM.pdf (accessed on 4 March 2020).

11. ResearchGate, (2012). *Soil Pollution Control Management Techniques and Methods*. [online] Available at: https://www.researchgate.net/publication/320395375_Soil_Pollution_Control_Management_Techniques_and_Methods/link/59e1cd01a6fdcc7154d7dbb4/download (accessed on 4 March 2020).
12. Root Zone Technology: Reviewing its Past and Present. (2015). [eBook] Available at: https://www.ijcmas.com/vol-4-7/A.%20A.%20Raval%20and%20P.%20B.%20Desai.pdf (accessed on 4 March 2020).
13. Soil and Groundwater Remediation Specialists, (n.d.). *Phytoremediation Technologies-Soil and Groundwater Remediation Specialists*. [online] Available at: https://www.xdd-llc.com/phytoremediation-technologies/ (accessed on 4 March 2020).
14. The Technologies for Active Noise Control, (n.d.). [eBook] Available at: https://www.jstage.jst.go.jp/article/ast1980/12/6/12_6_245/_pdf (accessed on 4 March 2020).
15. Usgs.gov. (n.d.). *What is Remote Sensing and What is it Used for?* [online] Available at: https://www.usgs.gov/faqs/what-remote-sensing-and-what-it-used?qt-news_science_products=0#qt-news_science_products (accessed on 4 March 2020).

INDEX

Printed in the United States
By Bookmasters